# Laser Micro-spectrochemical Analysis

# LASER MICRO-SPECTROCHEMICAL ANALYSIS

H. MOENKE
L. MOENKE-BLANKENBURG

*VEB Carl Zeiss, Jena*

*translated by*
R. AUERBACH

CRANE, RUSSAK & COMPANY, INC.
NEW YORK

This volume is a translation of
*Einführung in die Laser-Mikro-Emissionsspektralanalyse*, 2. Auflage,
first published in 1968 by
Akademische Verlagsgesellschaft Geest & Portig K.-G., Leipzig


Published by
ADAM HILGER
Rank Precision Industries Ltd
29 King Street, London, WC2E 8JH

Published in the United States by:
Crane, Russak & Company, Inc.
52 Vanderbilt Avenue
New York
N.Y. 10017
Library of Congress Catalog No. 72–82876
ISBN 0-8448-0060-0


Printed and bound in German Democratic Republic
by Offizin Andersen Nexö, Graphischer Großbetrieb, Leipzig BT I

## From the Preface to the First Edition

For any true micro-spectrochemical analysis a microscope and a dispersing instrument are needed. Microspectrophotometers, outfits consisting of an ultra-violet spectrophotometer and a microscope, mainly intended for absorption measurements on organic substances, have been commercially available for decades. A reflecting microscope that can be attached to various infra-red spectrophotometers also provides information on the type and the composition of inorganic and organic compounds through recordings of absorption spectra.

Laboratories equipped for emission spectroscopy in the ultra-violet, visible and near infra-red spectrum regions, however, have so far carried out so-called spot analyses with microscope and spectrograph used separately and not simultaneously. The sample as a rule had to undergo two time-consuming procedures of preparation.

The invention, in 1960, of the solid-state laser by Maiman made possible a new instrument assembly, essentially consisting of a ruby laser and a research microscope. If a microscope fitted for visual observation, measuring and irradiation of very small sample zones with time and space coherent laser radiation, is linked with a spectrograph, this gives the first really useful and practical outfit for emission micro-spectrochemical analysis in the wavelength range from 200 to 1000 nm. It is theoretically possible to concentrate a light beam with a laser probe on a sample so that a focal spot of extremely high energy densitiy and of a diameter of the order of magnitude of the wavelength of the radiation emitted by the resonator is formed on the surface.

The authors took charge of and tested the model LMA, the first laser micro-spectrum analyser produced by VEB Carl Zeiss, Jena, which at the 1965 Leipzig Spring Fair won a gold medal for outstanding quality. This book contains a comprehensive report of their experiences; it will be of interest to chemists, mineralogists, geologists, metallographers, silicate technicians, biologists and medical men, as well as archaeologists, art historians and criminologists.

The authors thank Prof. Dr P. Görlich for his interest in their investigations. They also express their gratitude to Dr K. Laqua, Dortmund; Prof. Dr M. Loquin, Paris; Prof. Dr C. W. Beck, Poughkeepsie; Dr J. W. Popow and Dr E. P. Markin, Moscow, and to representatives and members of the staff of Optics Technology Inc., Palo Alto, Cal., and Jarrell-Ash Co., Newtonville, U.S.A., for valuable discussions and for permission to publish their illustrations. Thanks are furthermore due to the staff of the Akademische Verlagsgesellschaft Geest & Portig K. - G., Leipzig, and of the printing works for their collaboration and the careful production of the book.

*Jena*
*Autumn 1965*

L. MOENKE-BLANKENBURG
H. MOENKE

## Preface to Second Edition

As a result of the general interest in the further developments in micro-spectrochemical analysis with a laser as energy source it became necessary only a year after publication to commence work on a second edition with its numerous additions and extensions. In particular, §§ 2.3, 2.4, 3.2 to 3.4 (chiefly § 3.4.2) and Chapter 4 (Applications) have been enlarged, taking into consideration the experiences gained in bringing the Zeiss laser micro-spectrum analyser LMA 1 into routine production. Special attention has been paid to information derived from the examination of a wide variety of sample materials from customers at home and abroad.

New solid-state lasers and accessories have been included in the survey in § 2.2. In Chapter 5, advances in X-ray micro-analysis are discussed in the comparison with other analytical methods.

Additions to the bibliography cover mainly publications of the years 1965 to 1967.

The authors express their thanks for valuable discussions to Prof. Dr H. Malissa and Prof. Dr E. Schroll, Vienna; Prof. Dr H. Kaiser, Dr Laqua and Dr Hagenah, Dortmund; Prof. Dr N. D. Topor, Mrs. Prof. Dr Schadlun and Prof. Dr Karyakin, Moscow; Dr E. G. Skworzov and Dr I. I. Kovner, Leningrad; Prof. Dr Dr h.c. P. Görlich, Ing. W. Sebastian, Dipl.-Ing. Wiegand and Dipl.-Ing. Birkner, Jena.

The authors are likewise grateful to the Akademische Verlagsgesellschaft Geest & Portig K.-G., Leipzig, and to the printing works for their kind collaboration.

*Jena*
*March 1967*

L. MOENKE-BLANKENBURG
H. MOENKE

## Author's Preface to English Edition

Laser micro-spectrum analysers are now being used in some thirty countries for spot analyses in research, routine examinations and industrial production control. This no doubt shows approval of the scientific instrument makers' well-considered initiative in linking optical microscopy and atomic spectroscopy, further developments of methods and applications resulting from gained experience all the time.

In contacts abroad the authors repeatedly met the call for translation of their book. In 1968 the first Russian edition[84] appeared in Moscow. The authors are very grateful to Adam Hilger Ltd, London, and to Akademische Verlagsgesellschaft Geest & Portig K.-G., Leipzig, for their helpful and welcome collaboration in bringing out this English translation.

The work has been extensively revised and enlarged to take into account developments of instrumentation and of working methods. Additions have been made in particular to §§ 1.1, 2.4, 3.1, 3.2, 3.3, 3.4.3.3, 4.1.2, 4.1.6 and 4.4. About fifty new references to recent publications have been included in the bibliography.

We are very grateful for all the care Mr R. Auerbach has taken in preparing his excellent translation of our book.

Thanks, herewith gladly acknowledged, are again due for fruitful discussions, namely to Dr J. Matsushita, Tokyo; Dr H. Sunahara, Osaka; Prof. Dr A. W. Karyakin, Moscow; Prof. Dr A. Danielson, Stockholm; Prof. Dr W. Eitel, Toledo, U.S.A.; Dipl.-Phys. H.-J. Mohr and Dipl.-Phys. W. Quillfeldt, Jena.

*Jena*
*January 1973*

L. Moenke-Blankenburg
H. Moenke

## Contents List

# 1

# Solid-state Pulsed Lasers as Energy Sources for Micro-analysis by Optical Emission Spectroscopy

## 1.1. PRINCIPLES OF THE LASER

In 1955 and 1956 the maser concept was adopted in microwave work for constructing novel type amplifiers and oscillators. The acronym MASER stands for Microwave Amplification by Stimulated Emission of Radiation.

Stimulated emission of radiation (Einstein 1917) requires an external electro-magnetic field that causes a particle to move from a higher to a lower energy level. This transition increases the external field strength because the energy given up by the particle in the transition is in phase. In 1958, after masers had been established, Schawlow and Townes[116] proposed the construction of infra-red and optical 'masers'.

This suggestion was readily taken up by many scientists and in 1960 Maiman[70] introduced a ruby LASER. This was the first 'Light Amplification by Stimulated Emission of Radiation' and also the first solid-state laser. As only solid-state lasers have been used in optical emission spectrum analysis it is not necessary here to discuss gas and semiconductor lasers.[109]

Certain basic conditions must be fulfilled for selective amplification by stimulated emission processes to occur in the closed system of a solid-state resonator:

1 The material of the resonator – the most important part of a solid-state laser – must have at least three discrete energy levels between which transitions are permitted.
2 An external electromagnetic field must exist, sufficiently strong to excite, through stimulated absorption, so many particles that the ground state becomes substantially depopulated.

To understand the general excitation process of the laser one has to distinguish the four discrete energy levels: ground level, terminal level, metastable level and pump level. This gives a four-level laser. In the three-level laser of Maiman the ground and terminal states are identical. The mechanism of optical pumping encompasses

1 The pumping from ground to the pumping level,
2 The transition, without emission, to the metastable level,
3 The stimulated emission ('signal transition') to the terminal level of the laser radiation, and
4 The transition, without radiation, to the ground level.

This excitation mechanism of optical pumping is necessary because, in a closed system of quantum mechanics, static equilibrium exists between excited and non-excited conditions, with a distribution given by the Boltzmann relation:

$$\frac{N_2}{N_1} = e^{-\Delta E/kT} \qquad (N_2 < N_1)$$

where $N_1$, $N_2$ = number of atoms in energy levels 1 and 2, respectively
$\Delta E$ = difference between energy levels
$k$ = Boltzmann constant ($k = 1{\cdot}38 \times 10^{-16}$ erg/degree)
$T$ = absolute temperature.

Therefore more atoms are at levels of lower energy and, consequently, resonance absorption is the rule in the optical range (including the infrared, visible and ultra-violet), that is intake of energy by non-excited atoms with subsequent spontaneous emission.

Stimulated emission requires the condition $N_2 > N_1$ (so-called population inversion). The $E_2$ level has to be metastable ($E_m$). Spontaneous emission then is not probable, the transition to the ground level being forbidden.

### 1.1.1. Solid-state resonators

Population inversion can be produced in every solid-state resonator. These resonators consist of host solids (monocrystals or glass) containing within the crystal lattice or the glass structure active additives of certain metal cations (activators) in concentrations of a few hundredths of a per cent up to several per cent. The resonator of the first solid-state laser was a cylindrical rod of synthetic ruby ($Al_2O_3$) doped with 0·05 per cent chromic oxide ($Cr_2O_3$). For optical pumping of a suitably arranged and prepared macroscopic resonator, a xenon flash tube is needed; the resonator and flash tube, together with a reflector, are housed in the laser head.

The cylindrical resonator rod of the first solid-state laser had reflecting coatings on its two plane parallel end surfaces, one fully and one partially reflecting. The flash tube, of helical form, surrounded the ruby rod. The plane reflecting surfaces formed a Fabry-Pérot interferometer whose dimensions were very large compared with the wavelength $\lambda$. The purpose of the interferometer is to increase the time of interaction between the spontaneously emitted radiation and the atoms of the activator. This leads to a greater number of excited atoms and through them to a higher energy supply to the wave field.

A Fabry-Pérot interferometer is in resonance if the distance between its two reflecting surfaces is equal to a whole number of half-waves ($\lambda/2$) of the light to be amplified. In the interaction between excited atoms and the lightwave, resonator energy transitions occur at resonator levels and an electromagnetic field can be generated. The build-up time of the field depends on the quality of the resonator rod. The field itself can produce amplification through stimulated emission and this finally leads to an overall increase of the stimulated emission, in the manner of a feedback system. The resonator oscillates by self-excitation when the energy take-up exceeds the losses.

The ruby crystal of the first Maiman laser was grown by the Verneuil method;[70] from it a resonator rod of 10 mm diameter was made.

If radiation of wavelength 550 ± 50 nm or 410 ± 50 nm is incident on such a solid-state resonator, transition from the ground level [$^4A_2$ doublet from doubly degenerate levels with 0·4 $cm^{-1}$ (12 GHz) separa-

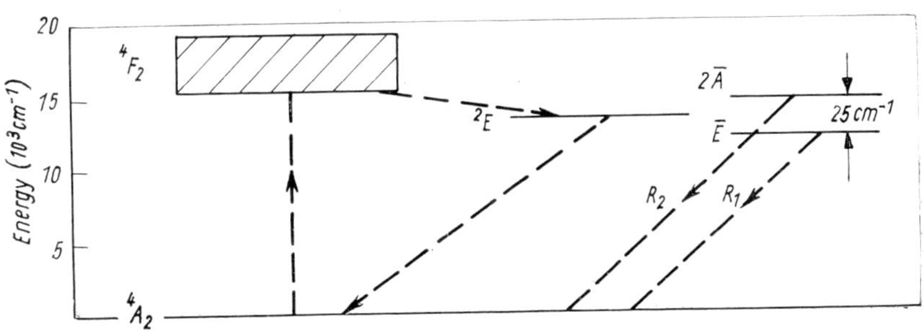


FIG. 1. Energy level diagram (part) of $Cr^{3+}$ ions in a ruby crystal.
Laser radiation: $R_1$ – 694·3 nm; $R_2$ – 692·9 nm.

tion, not resolved at 300 °K] into the two broad absorption bands $^4F_2$ and $^4F_1$ occurs (Fig. 1). From there the excited $Cr^{3+}$ ions within less than $10^{-7}$ s go over, without emission, into the two terms $^2E2\bar{A}$ and $^2E\bar{E}$ (separation 25 $cm^{-1}$; 870 GHz) of the metastable level $^2E$; this level is then depopulated in spontaneous emission in a sharp doublet (694·3 and 692·9 nm at 300 °K).

The ruby terms are very dependent on the operating temperature: at room temperature the red line $R_1$ is emitted at 694·3 nm with a width of 200 GHz; at 77 °K it lies at 693·4 nm with a band width of 15 GHz. Although mercury arcs and mercury vapour lamps are more suitable for pulsed as well as for continuous laser work with a ruby, the xenon flash tube is still widely used as pumping source.

In spite of the not very satisfactory energy aspect of the three-level laser, the ruby laser is still very much preferred because of the visibility of the radiation emitted mainly at 694·3 nm, and so is the simple arrangement of the Fabry-Pérot interferometer because only relatively few modes are simultaneously excited.

The resonator of a four-level laser, also in wide use, is made of glass doped with neodymium. Its operating temperature is 300 °K; the emission wavelength, 1060 nm, lies in the near infra-red. This material is made, amongst others, by the VEB Jenaer Glaswerk Schott & Gen.

In the neodymium laser the pump transition leads from the terminal state to several rather narrow bands which can be populated by excited $Nd^{3+}$ ions. These pump bands are ten times narrower than corresponding bands in the energy level scheme of the corundum lattice. Neodymium glass resonators are normally excited by xenon flash tubes. The pumping range is 500 to 800 nm. The optimum active ion concentration is about 6 per cent. For the different types of neodymium glasses the pump transition becomes narrower at low temperatures (77 °K) and the efficiency is then reduced.

After population of the upper laser term which leads to the metastable state with a spontaneous lifetime of about 0·7 ms the signal transition from there to the terminal state occurs:

$$^4F_{3/2} \rightarrow {}^4J_{11/2}$$

$^4J_{11/2}$ lies approximately 2000 $cm^{-1}$ above the ground state and is very sparingly populated at 300 °K. A population inversion, very little affected by a change in temperature, is obtained with remarkably low pumping. An advantage of neodymium resonators of large dimensions is the existence of pump bands in the red portion of the visible spectrum; their use allows considerable reduction of the thermal stress. Because the field surrounding the ions in neodymium glasses is very variable the emitted lines are relatively broad (about 30 nm) compared with $Nd^{3+}$ crystal resonators. The resulting lower amplification at comparable inversion is partly compensated by high doping rates and better homogeneity of the laser glasses.

The American Optical Company, Southbridge, developed the so-called AO Lux laser rods. Barium crown glass ($n_{1\cdot06\,\mu m} = 1{\cdot}509$) is doped with 5 per cent (mass) $Nd_2O_3$. The diameters of the resonators range from 0·125 inch (about 3 mm) to 2·75 inches (about 70 mm). Special rods have been made up to 2 m long.[4] The best output energy for this material is obtained when the length to diameter ratio is about 40. Rods of 3 inch (about 76 mm) length and 0·125 inch (about 3 mm) diameter have the remarkably low threshold value of 12 Ws. Technological experience gained in the manufacture of glass fibre optics led to the development of neodymium glass with a double sheath. These Clad AO Lux Laser Rods have an active neodymium barium crown glass core of two-thirds of the total outer diameter. The inner sheath consists of EOD 18 glass which absorbs fluorescent light particularly well. The outer protective layer is R6 glass. The double sheath prevents loss of coherent radiation because of its strong absorption at 1060 nm. The system also serves for focusing the pump radiation in the active core ($4{\cdot}6 \times 10^{20}$ $Nd^{3+}$ ions/$cm^3$).

Neodymium is also used for doping Ca and Sr tungstates and Ca, Sr and Pb molybdates. Apart from its presence in these anisodesmic structures, it is also found as activator in the lattices of $CaF_2$ and $LaF_2$. Resonators made from these materials emit radiations of wavelengths which are always above 1040 nm, the corresponding pump bands (absorption wavelengths) lying in the visible spectrum region between 500 and 800 nm.

Alkaline earth fluorides doped with $U^{3+}$ emit radiations of wavelengths above 2400 nm.

Rare earth ions too have been proved to be good activators. Several dozens of substances from which resonators for solid-state lasers may be made have so far been found. They have mainly been tested experimentally. The fundamental research in this special field of crystallography is far from finished; it calls for close collaboration between workers in crystal chemistry and spectroscopy.

The length of an optical resonator is very great compared with the wavelength, $\lambda$, of the radiation oscillating in it. Therefore a large number of eigenvibrations has to be expected.

The output from a Fabry-Pérot resonator which contains a medium with negative absorption coefficient for producing the laser effect is emitted at right angles to the two parallel reflecting surfaces: for, according to Fox and Li Tingye,[34] of the many possible modes or eigenvibrations only the axial $TEM_{pqr}$ modes of low transverse order $p$ and $q$ are excited. In this group of axial modes the dominant $p = q = 0$ is of specially high quality.

Once the reflecting surfaces of the Fabry-Pérot plates are no longer exactly parallel the output usually becomes diffuse.

### 1.1.2. The properties of laser radiation

The radiation emerging within a very small solid angle from a resonator with reflecting plane-parallel interferometer plates is time coherent (hence very purely monochromatic) – different from the incoherent light sources known until 1960; it is also spatially coherent – laser radiation therefore can be concentrated into a very narrow beam, either through appropriately shaping the resonator or by means of a suitable external optical system; and laser radiation has, as a consequence of the small bandwidth and of the focusing of the beam, a very high spectral energy density.

### 1.1.3. Efficiency, power, radiation density and energy of solid-state pulsed lasers

Most commercially produced ruby lasers are excited with a xenon flash tube and are pulse operated. A laser consists of two parts: the laser head and the electronic mains supply unit. The largest component of the latter is the capacitor bank required for driving the flash tube. The performance characteristics of the exciting system of interest for comparison purposes are the capacitance, the capacitor charging voltage and the energy of the flash tube pulse ('input' of the laser; for a medium-sized laser of the order of several thousands joules).

The efficiency of the solid-state lasers known today is very low, about 0·001 per cent up to 3 per cent at the most. Instruments with solid-state lasers, consuming an input energy of several thousands of joules, emit about 2 to 5 J.

An output of 1500 J needed a flash tube energy of $12 \times 10^4$ J; the capacitor bank had a mass of 4·5 metric tons.[122] Such equipment cannot, of course, be considered as a component of routine equipment for micro-spectrochemical analysis. In spite of their low efficiency the output of commercial lasers with values of several hundreds up to a few thousands watts is relatively high. It exceeds by several orders of magnitude the corresponding values for incoherent light sources.

In the operation of a normal pulsed laser the population of excited levels is reduced to a density below the threshold for stimulated emission in about 0·5 ms; after completion of the process sufficient excited states per unit volume no longer remain. After adequate cooling of the resonator the flash tube therefore has to be operated again.

If the lifetime of an atom in the excited state is $\tau$ seconds, $N$ excited atoms emit $N/\tau$ wavetrains per second of which only those parallel to the axis are of importance for the laser process in a Fabry-Pérot system, their number increasing with increasing wavelength. The number of laser waves per second and per cubic centimetre, developing spontaneously, is

$$N_{\mathrm{L}} = \frac{\lambda^2 c}{8\pi^2 \Delta\nu} \frac{N}{\tau}$$

where $c$ = the velocity of light
$\Delta\nu$ = frequency spread.

The number of the stimulated light quanta is given by multiplying this equation by $n_{\mathrm{L}}$, the number of the light quanta already existing. It is a necessary condition of laser action that the increase per second of these quanta exceed the losses (for instance through escape through the reflecting surface), which means that the increase must be greater than $n_{\mathrm{L}}/t$ ($t$ being the time in seconds). $N_{\mathrm{L}}$ no longer appears in the laser condition if it is expressed by the inequality

$$\frac{\lambda^2 c}{8\pi^2 \Delta\nu} \frac{N}{\tau} > \frac{1}{t}$$

This relationship leads to the following conclusions: To produce a large excess value of $N$ the resonator has to be irradiated with light of short wavelength. Thereby the electrons of the activator are raised to high energy levels from which they can go over to the initial level of the laser action proper. The left-hand term of the inequality should have as high a value as possible. Therefore activators always have atoms with small frequency spread $\Delta\nu$. A high value of $\lambda$, that is a long wavelength, means that laser action is more frequently produced in the infra-red than the visible spectrum region. To reduce the value of $1/t$ as much as possible, high reflectivity is used (multilayer dielectric coatings have superseded the earlier silver, gold or other metal coatings) and the separation of the reflectors is made large compared with $\lambda$.

In micro-spectrochemical analysis with laser excited emission the test sample is placed in the focal plane of a microscope objective. It should be pointed out that the laser has advantages for this sort of vaporization. A non-coherent light source gives at the focus of a simple, single lens objective a maximum density of $10^3$ W/cm$^2$. The coherent radiation of a solid-state laser with a pulse power of more than $10^3$ W, produces at the focus of the same lens an energy density of at least $10^{12}$ W/cm$^2$.

## 1.1.4. Modified lasers

### *1.1.4.1. Pump sources and reflectors*

The first xenon flash tubes were helix shaped. Then straight tubes were introduced which called for different reflectors; resonator and tubular flash tubes were often arranged at the conjugate lines of a reflector in the shape of an elliptical cylinder.

In the Siemens solid-state lasers, mercury capillary and quartz iodine lamps are used besides xenon flash and arc tubes; the tubes and the resonator are arranged in a line one behind the other on the axis of rotation of an ellipsoidal mirror. The improvement in the illumination arrangements and the symmetry of the pumping beam resulted in an increase of the efficiency of the ruby laser from 0·5 per cent to 1·5 per cent. This system made it possible to run ruby, YAG, $CaF_2/Dy^{2+}$ and glass/$Nd^{3+}$ lasers continuously at room temperature.

High power lasers have several flash tubes appropriately grouped in the laser head round the resonator. In one possible arrangement the tubular xenon tubes are placed at the focal lines of a quadruple ellipsoid with the ruby crystal rod at the common centre of symmetry of the assembly.

Glass works at Mainz have developed a so-called focusing cylinder of

Duran 50 glass for the purpose of reducing the thermal stress in the resonator. The ruby and flash tube are in the positions of the focal lines, 28·0 or 30·5 mm apart, of the ellipsoidal cylindrical reflector with a multilayer dielectric coating. The coating transmits the unwanted light and only reflects the radiation of the pumping frequency which is concentrated on to the ruby rod at the focal line.

The Deutsche Glimmlampen-Gesellschaft Pressler at Leipzig has for some time been making helical and straight high-power electron flash tubes of quartz which are being used for laser pumping (Zeiss ZFL Instruments). So far four lamps in particular (see Table 1) have found application in laser micro-spectrochemical analysis.

TABLE 1. Xenon flash tubes of the Deutsche Glimmlampen-Gesellschaft Pressler, Leipzig

| *Catalogue number* | *Shape* | *Maximum running voltage* $U_{a\,max}$ (kV) | *Minimum running voltage* $U_{a\,min}$ (kV) | *Maximum load* $A_{max}$ (J) | *Spontaneous ignition voltage* $U_Z$ (kV) |
|---|---|---|---|---|---|
| G 431 | Helical | 3·2 | 1·5 | 750 | ≥3·5 |
| G 455 | Straight | 2·0 | 0·8 | 1000 with $L = 100\,\mu H$ | ≥ 3 |
| G 453 | Helical | 3·0 | 2·0 | 2500 | ≥3·5 |
| G 473 | Straight | 2·0 | 0·75 | 1800 with $L = 100\,\mu H$ | ≥3·0 |

In 1967, tube type G 455 was superseded by type G 495 which has internal reinforced electrodes and a lifetime two or three times as long.

In the USSR helical as well as straight flash tubes are used for pumping solid-state lasers. The quantum generator GOR-100 has surrounding its neodymium glass resonator a helical flash tube, ISKCH 5/30-14, which takes a maximum load of 12 kJ. A Q-switched ruby laser of 50 MW pulse power sits in a double ellipsoid reflector with two straight lamps. The optical quantum generator GSU-1 has in the resonator housing a rectangular block of neodymium glass and eight flash tubes of the type PFK-2000 providing a maximum pump energy of 15 kJ.

### *1.1.4.2. The resonator*

The pink ruby crystal used in 1960 by Maiman of the Hughes Aircraft Company, Newport Beach, USA, contained 0·05 per cent $Cr_2O_3$. In 1961 Varian BTL brought out ruby resonators with a $Cr_2O_3$ content of 0·5 per cent. Measurements showed the working temperature of these resonators to be below 78 °K and the wavelengths of the emitted laser radiation 700·9 and 704·1 nm.[36]

At the end of 1962 the Adolf Meller Company already listed eleven differently doped ruby types (0·006 to 1·2 per cent Cr) with three basically different crystallographic orientations.

Continuous operation of a solid-state laser at low temperatures of about 78 °K was achieved in 1962 with rubies having as activator essentially a chromium content of 0·5 per cent or less.

In 1965 Siemens brought out quasi-continuous solid-state lasers, working at room temperature and having ruby neodymium glass, $CaWO_4/Nd^{3+}$ or yttrium-aluminium garnet resonators.

Solid-state laser rods have to be worked to highest precision. With crystalline substances, adjustment and orientation of the synthetic raw materials have to be carefully controlled. The trigonal symmetry axis must as a rule be parallel to the cylinder axis of the resonator rod at least within a few minutes of arc. The plane end surfaces of the cylinder must be polished flat to a tenth of a wavelength of the sodium D line and parallel to less than 6 seconds of arc. Efforts have been made in recent years to make the multilayer dielectric coatings of the surfaces forming the interferometer as homogeneous and hard wearing as possible.

Flatness and parallelism of the resonator surfaces affect the pumping energy required for obtaining laser action. Attempts to reduce the diffraction losses at the edges of Fabry-Pérot systems led to the construction of resonators with confocal spherical surfaces which emit the laser radiation in spherical waves. An additional lens system has then to be used if a parallel beam of coherent light is to be obtained.

For laser micro-spectroscopy oriented resonators are needed; ruby crystals in particular have to be grown very slowly. So-called 90° rods are recommended for air-cooled lasers like those of all laser microprobe outfits so far commercially available: in these rods the optical axis of the crystal is at right angles to the cylinder axis, as in rods of the Hughes Aircraft Corporation, Newport Beach, California, which are about 9·5 mm (0·375 inch) in diameter and about 76 mm (3 inches) in length, with hard $MgF_2$ coatings.

Ruby rods with 0° orientation, that is with the trigonal crystal axis coincident with the cylinder axis, are mainly used at temperatures of about 80° K. The available maximum length is restricted by the crystal configuration.

The high-quality, slowly grown 0° ruby, Type 222, made by the Maser Optics Inc., Boston, Mass., of about 51 mm (2 inch) length, and 6 mm (0·25 inch) diameter, with plane end surface flatness of one-tenth wavelength and parallelism of 6 ± 3 seconds of arc, has a threshold value of about 160 J; it is used in investigations carried out with non-polarized coherent light.

The 90° ruby laser rod, Type 223, produced by the same makers is worked to the same precision, with the cylinder axis normal to the optical axis of the crystal to within 1°; it has a lower threshold value of 130 J and emits polarized coherent laser light. Jeol of Tokyo supply ruby rods with 60° orientation and also rods up to 20 mm diameter and 300 mm length.

Resonators with random axis orientation, as for instance the Type 221 standard ruby laser crystal for the light-weight system 'Series 600' of Maser Optics Inc., Boston, Mass., are only suitable for demonstration of the laser effect and similar simple experiments in teaching.

In the USA ruby rods are now being produced for standard laser equipment with a diameter of about 6 mm (0·25 inch) and lengths of about 25 mm, 51 mm, 64 mm and 76 mm (1 in., 2 in., 2·5 in. and 3 in.), for instance by:

Isomet Corporation, Palisades Park, N.J.
Valpey Crystal Corporation, Holliston, Mass.
Perkin-Elmer Corporation, Norwalk, Conn.
The Perkin-Elmer Corporation also offers resonators of approximate dimensions 9·5 mm × 152 mm (0·375 in. × 6 in.).
Isomet and Valpey have special rods up to about 228 mm (9 in.) long.
Meller had already made 305 mm (12 in.) long rods in 1962.

The coatings of the resonator surfaces are either of silver, as those of the Isomet lasers, with a transmission of the exit surface of 2 per cent or 4 per cent, the other surface being opaque with a gold overcoating, which facilitates alignment considerably; or they are multilayer dielectric systems, consisting of a number of very thin layers of alternating high and low refractive index. Such coatings can give a reflectivity as high as 99·9 per cent. The Isomet rubies with dielectric multilayer coatings have on the output side transmissions of 4 per cent, 10 per cent or 15 per cent and of less than 1 per cent at the other end.

A special version are the Meller and Valpey rubies with sapphire sheathing; this serves for focusing the pumping radiation into the especially active axial region of the crystal resonator.

Even in standard ruby resonator rods (without sapphire sheathing) the effective diameter $d_e$ is not the same as the overall external diameter $d$. The relationship between these two quantities is

$$d = d_e n$$

$n$ being the refractive index. For ruby this becomes

$$d_e = \frac{d}{1{\cdot}75}$$

Often the diameter of the resonator is smaller than the entrance pupil of the microscope objective. If a focal spot of very small diameter is to be

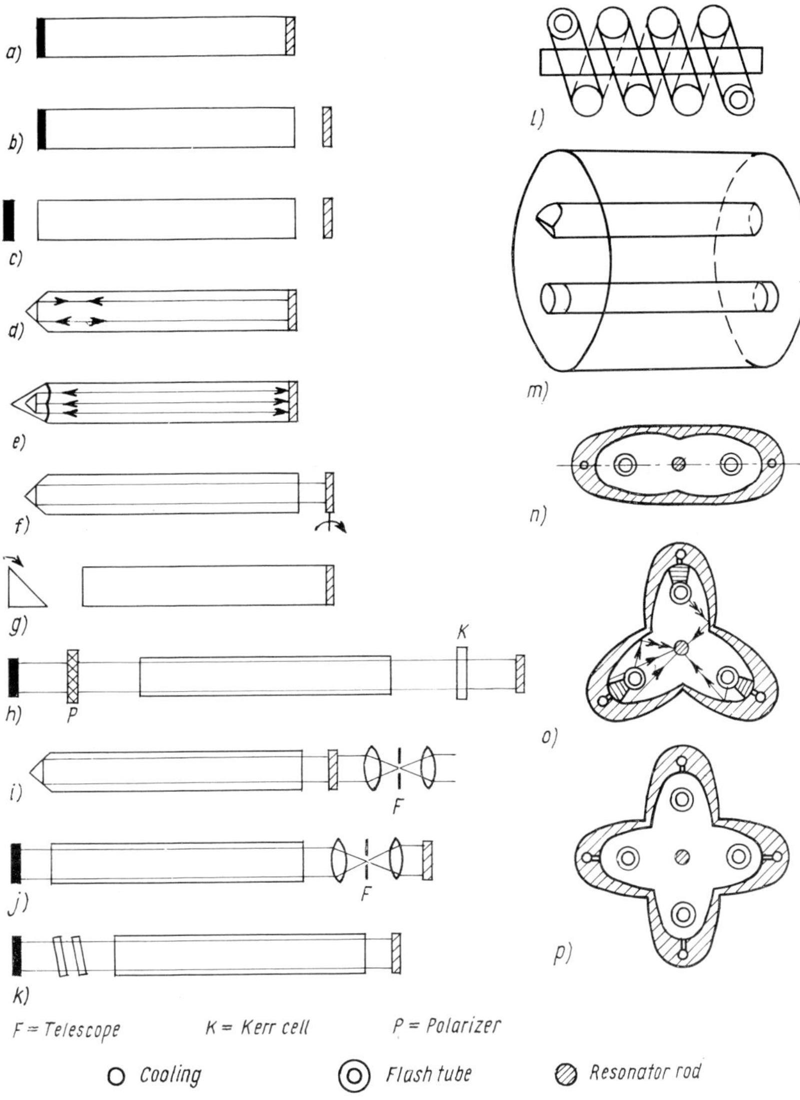


FIG. 2. Different types of resonator mirrors, of divergence reduction and of pumping sources and resonator assemblies.

*(a)* Standard Fabry–Pérot interferomerer with partially transmitting coating on output side. *(b)* External partially transmitting interferomerer plate. *(c)* Both interferometer plates external. *(d)* Totally reflecting roof prism. *(e)* Totally reflecting corner cube. *(f)* Q-switch mode interferometer (interferometer output place acting as rotating mirror). *(g)* Q-switching with rotation prism. *(h)* Q-switching with Kerr cell. *(i)* Telescope system with divergence reducing diaphragm between interferometer output plate and objective (not shown). *(k)* Intermediate imaging system for divergence reduction between resonator rod and external interferometer output plate. *(l)* Divergence reduction through a second Fabry-Pérot interferometer. *(m)* Illumination system: helical xenon flash tube, resonator with plane-parallel end surfaces. *(n)* Illumination system: straight flash tube, resonator with totally reflecting roof prism in elliptical reflector. *(o)* Illumination system: double elliptical reflector with two straight pumplight sources (Raytheon, Waltham U.S.A.). *(p)* Illumination system: treble elliptical reflector with three straight flash tubes. *(q)* Illumination system: quadruple elllptical reflector with four straight lamps.

formed on the sample, the laser head is placed in front of the eyepiece of a compound microscope; in this way the available part of the laser radiation is limited by the eyepiece diaphragm. The diameter of the diaphragm is

$$d_0 = \frac{50}{\Gamma} A$$

where $\Gamma$ is the total magnification and $A$ the numerical aperture of the objective.

If with a ruby rod of $d_e$ value 1·5 mm (that is $d$ about 3 mm) an objective of magnification ×100 and numerical aperture 1·25* is used, $\Gamma$ must be less than 400. For an eyepiece magnification of about ×4 the diameter of the focal spot is then a few micrometres.[5]

After resonators with two plane-parallel end surfaces (Fabry-Pérot interferometer systems) further forms have been developed. For instance the totally reflecting end can be constructed in some prism form: totally reflecting chisel ends (roof prism) and hexagonal facet ends are usually chosen (Fig. 2).

$Nd^{3+}$ doped $CaWO_4$ resonators give, for the same flash tube energy, an output three to four times as great as that of rubies. The most important solid-state resonators for continuous laser operation are:

$CaWO_4/Nd^{3+}$ ($\lambda$ = 1060 nm)

$SrMoO_4/Nd^{3+}$

$CaF_2/Dy^{2+}$

$CaF_2/U^{3+}$

$CaF_2/Dy^{2+}$ is remarkable for its very low threshold value, which is for instance only about one eighth of that of $CaWO_4/Nd^{3+}$.

Yttrium aluminium garnet is a new resonator material. The Korad Corporation, Santa Monica, California, has developed a laser of this material with an energy output of 0·5 J at a wavelength of 1060 nm. It has the advantage of very small mass. Further advantages of the material, a cubic crystal, are its optical homogeneity, which is better than that of other $Nd^{3+}$ crystal resonators, and the double doping with $Cr^{3+}$ and $Nd^{3+}$ ions. The latter makes possible an effective resonance absorption process, in which the red pumping bands of the $Nd^{3+}$ ions (4-level process) are populated by laser radiation of the $Cr^{3+}$ ions (3-level process). The spontaneous emission of the $Cr^{3+}$ ions can be initiated through broad band pumping with a mercury high pressure lamp.

---

* This specification will in the following be denoted in the abbreviated form: ×100/1·25.

YAG/$Nd^{3+}$ rods, 30, 55 and 80 mm in length, are used in the quasi-continuous solid-state lasers of the Siemens model A 2. The continuous Raytheon laser LCW 3 (0·5 to 0·75 W) also has a garnet resonator, whereas $CaWO_4/Nd^{3+}$ is the material of models LCW 1 and LCW 2, emitting 25 mW and 0·5 W respectively.

It has to be noted that the highest possible output energy can only be obtained with selected resonators and with adequate cooling of the laser head. The nominal energy values depend on the reflectivity of the surface coating or of the surfaces of the prismatic rod ends.

Ruby resonators have, owing to existing off-axis oscillation modes, a divergence of about $10^{-2}$ radian. Neodymium glass resonators produce less divergence because they can be made much more homogeneous.

TABLE 2. Solid-state resonators

| *System* | *Laser wavelength* (nm) | *Working temperature* (°K) |
|---|---|---|
| Glass/$Gd^{3+}$ | 310 | 77 |
| Plastic/$Eu^{3+}$ | 610 | 77 |
| $Al_2O_3/Cr^{3+}$ (Ruby)* | 694·3 | 300; 77 (Cont.) |
| $CaF_2/Sm^{2+}$ | 708·2 | 77 |
| $SrF_2/Sm^{3+}$ | 710 | 77 |
| Glass/$Yb^{3+}$ | 1010 | 77 |
| $CaF_2/Nd^{3+}$ | 1046 | 300 |
| $CaWO_4/Pr^{3+}$ | 1050 | 77 |
| $PbMoO_4/Nd^{3+}$ | 1058·6 | 300 |
| Glass/$Nd^{3+}$ * | 1060 | 300 |
| $BaF_2/Nd^{3+}$ | 1060 | 300 |
| $SrF_2/Nd^{3+}$ | 1060 | 300 |
| $LaF_3/Nd^{3+}$ | 1060 | 300 |
| $NaLaMoO_4/Nd^{3+}$ | 1060 | 300 |
| $SrMoO_4/Nd^{3+}$ | 1064·3 | 300 |
| | 1064 | 77 (Cont.) |
| $CaWO_4/Nd^{3+}$ * | 1064·6 | 300 |
| | 1063 | |
| $CaMoO_4/Nd^{3+}$ | 1067 | 300 |
| $CaF_2/Tm^{2+}$ | 1116 | 4 |
| $CaWO_4/Tm^{3+}$ | 1116 | 77 |
| $CaWO_4/Er^{3+}$ | 1612 | 77 |
| $SrF_2/Tm^{3+}$ | 1910 | 300 |
| $CaWO_4/Ho^{3+}$ | 2046 | 77 |
| $CaF_2/Ho^{3+}$ | 2050 | 77 |
| $CaF_2/Dy^{2+}$ * | 2360 | 77 (Cont.) |
| $SrF_2/U^{3+}$ | 2407 | 77 |
| $CaF_2/U^{3+}$ | 2556 | 77 (Cont.) |
| $BaF_2/U^{3+}$ | 2556 | 77 |

* These materials are of special importance.

To obtain greater energy densities or smaller light patches through focusing, the angle of divergence $\sigma$ has to be reduced because of the relationship:

$$\text{Light patch diameter (in focal plane) } x = 2f\sigma$$

$f$ being the focal length of the microscope objective. The divergence is given by:

$$\sigma \sim \frac{1{\cdot}2\lambda}{d_e}$$

where $d_e$ = effective diameter of the radiator
$\lambda$ = wavelength of emitted radiation.

This reveals that the theoretical calculated angle of divergence ($10^{-4}$ radian) is smaller by two orders of magnitude than the one experimentally obtained.

In 1964 Kim and Namba[61,92] reported success in experiments to reduce the divergence by using an external output mirror, coated for partial reflection with transmission of 2 per cent. Calorimetric measurements showed that the angle of divergence $\sigma$ decreases with increasing separation (0 to 600 mm) of the uncoated exit surface of the ruby (diameter 6·3 mm, length 63 mm) and the external reflector. The output energy of the resonator is, approximately, inversely proportional to the square of the angle of divergence. Therefore this method of mode selection directly only yields a reduction of the focal spot size; no greater energy density results.

Ruby crystals grown by the Verneuil process show some undesirable structural defects which can affect the oscillation modes and cause energy losses. According to Zaminer [Report at DGG/DKG (German Society of Glass Technology/German Ceramics Society) Conference at Würzburg in 1964] these defects develop during growth as well as in cooling: in the first case they consist in layer formation to the $11\bar{2}0$ planes and the 0001 planes; and in the second case in plastic crystal deformation in the $11\bar{2}0$ planes, anomalous optical bi-axial configuration due to elastic deformation; and tension stress in the outer crystal layers, compression stress in the core. Subsequent annealing can lead to decomposition of the crystal into polygones, the orientations of which may differ by several angular degrees.

What we have said so far will suffice to convey an idea of the difficulty of producing resonator crystals of highest quality.

### *1.1.4.3. Q-switching technique*[88]

The human eye can only perceive one light flash in a single laser pulse discharge. The energy content of a resonator, however, is discharged in a series of several dozen oscillating vibrations, that is in a statistical sequence of 'spikes' of different half bandwidths and different amplitudes. This fact can fairly easily be demonstrated by means of a suitable oscillograph. In 1962 McClung and Hellwarth[76] suggested controlling the performance of laser resonators, that is 'switching' them so that well-defined pulses, or only one pulse for each pumping action, are given off.

If, by a light switch, the amplification or the resonator performance is kept low until a large overpopulation is built up in the active substance, giant pulses result. In these, population inversion is restored in a very short time interval (less than $10^{-7}$ s).

Generally one distinguishes between active and passive switching. Additional energy is needed for active switching (voltage for opening the switch, power for rotating reflectors). Passive switching does not require additional energy. It is carried out with an absorbing material the transmission of which depends on the light flux in the resonator.

Electro-optical devices such as Pockels cells or Kerr cells combin edwith a polarizer in the resonator unit are rapid active switches. Considerable auxiliary equipment however is needed. The high voltages necessary for the switching have to be applied within a very short time interval. As polarizers, prisms of the Nicol, Wollaston or Glan-Thompson type with uncemented surfaces ('air-spaced') may be used, but only for moderately high energies. Stacks of plates at the Brewster angle are preferable as polarizers for the highest pulse energies. Systems for rotating the plane of polarization, which incorporate electro-optical crystals such as KDP (potassium dihydrogen phosphate) or ADP (ammonium dihydrogen phosphate), have the advantage that the control voltage for a 90° rotation in a double pass is relatively low, namely about 5–12 kV. But such crystal cells are rarely used because they are ruined by the laser radiation. In experiments with large rubies pulse durations of 10 ns and peak outputs of several hundred megawatts have been obtained. Kerr cells filled with nitrobenzene are used more frequently because they take a greater load and are easier to handle. Their control voltage is 20–30 kV. The nitrobenzene of the cell must be extremely pure. At high power, stimulated Raman scattering can give rise to satellite lines which use up energy. Pulses of 50 MW and of 20–30 ns lifetime are to be expected.

Another method of Q-switching is that of rotating reflectors. Such a reflector takes the place of one of the plates of the Fabry–Pérot system. With the reflector in the appropriate position the resonator starts to oscillate when the population inversion in the active material is sufficient. The pumping therefore must begin before the resonator reaches its optimum condition. The discharge of the flash tube is electronically synchronized in such a way with the parallel orientation of the prism or mirror that the flash occurs at the moment of maximum population inversion. When the reflector is in its parallel position the excited levels are depopulated and the photon density in the resonator increases; it reaches its maximum when the inversion is restored. The resonator condition deteriorates again as the prism (or the reflector) goes on rotating; the continuing excitation process causes a new population inversion to build up, but the threshold is not reached again. These are the conditions only near the threshold. With pronounced excess pumping or with not fast enough rotation of the reflector multiple pulses may occur. The correct interval between the flash tube discharge and the moment of parallel orientation of the reflector has to be determined experimentally.

The optical delay depends on the quality of the active material, on the lifetime of the excited conditions, on the change of the exciting energy with time and on the resonator geometry. The power consumption is less than that of electro-optical switching.

With synchronous drives the first single pulses are obtained at 10000 r.p.m.; their lifetime does not appreciably decrease with increase of the speed of rotation up to 24000 r.p.m. With turbines 100000 r.p.m. can be obtained, with special helium driven turbines up to 200000 r.p.m. Turbines however can only drive small mirrors and their angular velocity often is not constant so that as a rule a simple fast electromotor is to be preferred.

With a rotating prism the energy decreases with increasing speed of rotation. Even small ruby lasers with rotating mirrors yield several MW in pulses of 50–100 mJ energy and 50–100 ns duration. The pulse power is approximately 20 to 50 times as high as that of the relaxation spikes in normal laser use with the same pumping power.

A resonator with two plane reflecting surfaces has to be aligned very accurately; therefore only resonators with one 90° prism and one plane mirror are used. If the axis of rotation is at right angles to the edge of the 90° angle of the prism only a coarse adjustment is required. The hypotenuse surface should have an anti-reflection coating to prevent formation of a second beam. Mechanical vibration and noise of the motor are troublesome in Q-switching with rotating reflectors.

Q-switching with saturable absorbers[123] is a simple and inexpensive but nevertheless satisfactory method, especially for laser micro-spectrochemical analysis. Suitable saturable absorbers are solutions of certain organic dyestuffs in low concentrations of $10^{-6}$ to $10^{-5}$ g/cm$^3$; placed in cells in the resonator unit they act as passive reversible switches and are not destroyed by the high laser power. The transmission of the dye must be chosen so that the maximum inversion just reaches the threshold of the complete system. Through the effect of the growing number of light quanta the transmission is stepped up within a few nanoseconds to nearly 100 per cent. Such Q-switching works extremely fast and is practically free from losses.[109]

Various types of cyanines can be used as dyes such as crypto- or rubrocyanine (bis-(1-ethylquinolin-4)-trimethine cyanine iodide). Cyanine dyes do not react with reducing agents, but are destroyed by acids, alkalines and oxidizing agents. They are sensitive to ultra-violet radiation (therefore no quartz or fused silica cells should be used), but are not affected by radiation of the wavelength 694 nm.

Q-switching with dyed foils and glass absorption filters did not give results comparing favourably with those of the dyestuff solutions, especially not with regard to reproductibility. Dyed foils for this application are offered by Beckmann and Whitley, San Carlos, USA, but could not be tested by the authors.

Q-switching with a combination of active and passive devices always ensures reproducible single pulse working with high output and extremely small half bandwidth values. Such performance characteristics, however, are not called for in laser micro-spectrochemical analysis; here maximum energy (about 0·3 J) is required with comparatively broad spikes and a medium peak output (about 6 MW is sufficient) to vaporize maximum sample quantity; excessive wear of the instrument parts along the radiation path should, however, be prevented.

Passive Q-switch devices are commercially available for instance from Maser Optics Inc., Boston, Mass., USA. One model of the laser microscope of this firm incorporates such a switch with the following specification: output energy 0·15–0·25 J; pulse lifetime less than 30 ns with peak power output of 5–10 MW, or 80–100 ns with 0·5–1·0 MW (prism rotating at 20000 r.p.m.), or 45–50 ns with 5–10 MW (prism rotating at 90000 r.p.m.). Compare this with the 2–4 J output energy of the non-switched laser (500 μs, 0·01 MW).

# 2

# Optical and Electronic Instruments and their Assembly for Laser Micro-spectrochemical Analysis

## 2.1. GENERAL EQUIPMENT

The basic units of an outfit for laser micro-spectrochemical analysis are a solid-state laser (laser head and supply unit), a microscope for use with reflected light, mounted on the optical bench of a spectrograph, and a condenser system of lenses, lens combinations or mirrors, set up between the microscope and the entrance slit of the spectrograph.

The laser head is attached to the microscope, usually fitted with a special electrode holder for fixing and adjusting the pair of auxiliary electrodes between the sample surface and front lens or protecting plate of the microscope objective. The electrical equipment comprises a mains supply unit which also serves for setting the spark gap parameters (electrode voltage, capacitance and inductance). Usually the spark discharge forms the actual light source for the spectrum analysis.

The spectroscopic unit is usually a photographic prism or grating spectrograph. For particular research projects direct reading spectrometers with photoelectric detection by photomultipliers have also been used.

## 2.2. LASERS FOR MICRO-SPECTROCHEMICAL ANALYSIS

The first outfit made by VEB Carl Zeiss, Jena (Fig. 3a) for laser micro-spectrochemical analysis incorporated the solid-state laser model ZFL 750, originally shown at the Leipzig Fair in Spring 1964. This and the Zeiss gas laser ZGL 900-V were the first commercially available lasers in the G.D.R.[12,129a]

The mains supply unit ZFL 750-T5 (dimensions 350 × 430 × 640 mm, weight about 80 kg) contains a bank of electrolytic capacitors, the charging unit and a trigger. The flash tube voltage can be continuously varied between 1 and 3 kV, corresponding to an energy varying between 85 and 750 J. In use of the laser for micro-spectroscopy the ignition of the flash tube is started by hand by means of a switch in the front panel of the supply console. For other applications of the laser a starter for automatic ignition of the flash tube, continuously variable between 5 and 20 s, is built in. Alternatively the flash tube can also be ignited through external voltage pulses between 3 and 100 V.

The laser head of the first Zeiss solid-state laser is trapezium shaped (140 × 120 × 180 mm). It contains the resonator (45–60 mm in length and 3–7 mm diameter), in the flash tube, a metal foil reflector and the ignition coil. The resonator is surrounded by the helical flash tube and in continuous use can be cooled by a blower (about 10 l/s).

The standard model ZFL 750 is mounted by means of a carriage on a triangular optical bench; an adjustable holder and arm for fitting the laser head to a supporting bar has been designed for use in micro-spectroscopy and other microscopy applications.

The first successful experiments in micro-spectroscopy with the model LMA, demonstrated at the Leipzig Spring Fair in 1965, were carried out with neodymium glass resonators (5 mm diameter, 60 mm length) and with ruby rods (5 mm diameter, 45 mm length) at the laser wavelengths of 1060 nm and 694·3 nm, respectively.

Zeiss, Jena, also offer for special work various alkaline earth fluoride resonators, such as $CaF_2$ with $U^{3+}$ ($\lambda$ 2556 nm), with $Nd^{3+}$ ($\lambda$ 1046 nm), or with $Sm^{2+}$ ($\lambda$ 708·2 nm); and $BaF_3$ with $U^{3+}$ ($\lambda$ 2556 nm) or with $Nd^{3+}$ ($\lambda$ 1060 nm). They guarantee the following precision: the end surfaces flat to less than one tenth wavelength of the sodium D line, parallel to less than 6 seconds of arc and normal to the geometrical cylinder axis within less than 10 minutes of arc; diameter ±0·1 mm and, with flat end surfaces, length ±0·1 mm of the nominal values.

The Zeiss resonators have either multilayer dielectric or metallic coatings with a reflectivity in the red spectrum region of between 90 per cent and 100 per cent.

The prototype LMA model incorporating the laser type ZFL 750 is an instrument with a fixed Fabry–Pérot interferometer. It was the first apparatus for laser micro-analysis of this type to become commercially available.[13,82] The maximum emitted laser energy is 0·2 J. In many cases, however, half this value is adequate for spectrum analysis.

The laser ZFL 750 can be run from a mains supply of 220 V, 50 Hz and takes, on average, for 10 flashes per minute a current of 2 A. Such a flash repetition rate, however, is not necessary for micro-spectroscopy, for which often a single laser burst is sufficient. When a sample surface is to be irradiated several times in succession with the LMA outfit it is advisable

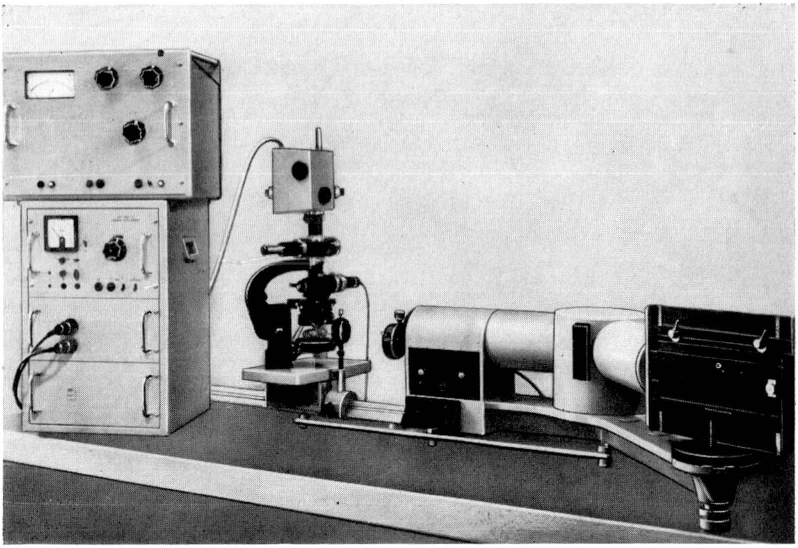

FIG. 3. The laser micro-spectrum analyser of VEB Carl Zeiss, Jena.

(*a*) Outfit for laser micro-spectrochemical analysis LMA (1964). From right to left: UV spectrograph Q 24, microscope with resonator head on optical bench, mains supply units for solid-state laser ZFL 750 and (on top left) for spark gap.

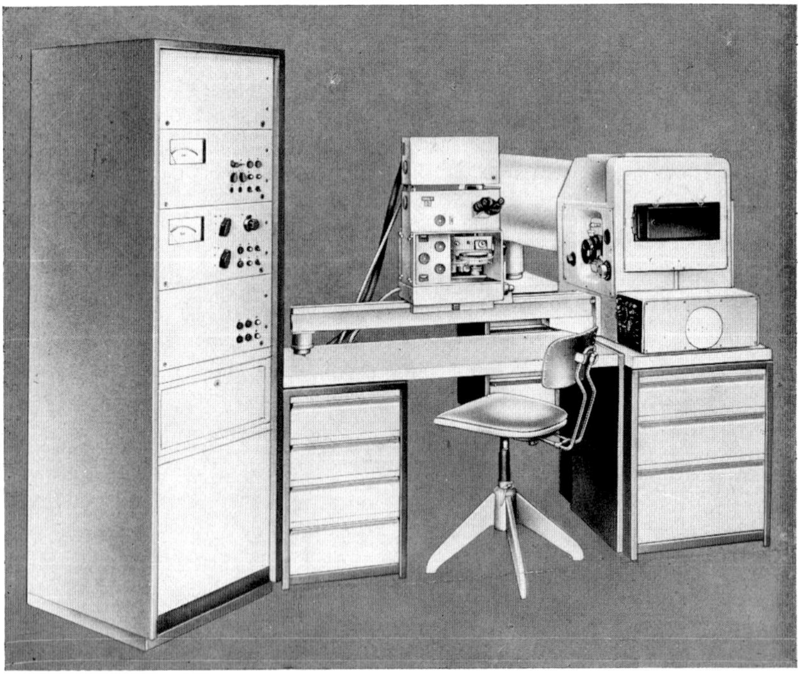

(*b*) Laser micro-spectrum analyser LMA 1 (1965). From right to left: 2 m Ebert plane grating spectrograph, microscope unit with laser head ZFL 1000 on optical bench, electronic console for supply of solid-state laser and of spark gap for cross excitation. (Top chassis for Q-switch working mode.)

to let the resonator cool off a few minutes in order to obtain maximum or near maximum output.

Two Italian outfits for laser spot analysis have also been designed without Q-switching (CISE Segrate, Milano).[25,128] The specification of the mains supply unit is: 320 μF, 0–2·5 kV continuously variable, pumping energy 100 J. The air-cooled ruby (doping 0·052 per cent $Cr_2O_3$, diameter 6 mm) of the model CISE I was later superseded by neodymium glass (doping 3 per cent, diameter 9 mm). The resonance reflector consisting of two sapphire plates, flat to one tenth wavelength with a reflectivity of 60 per cent, has been retained. The ruby outfit yielded, with a flash tube voltage of 1·525 kV, an output of $1{\cdot}5 \times 10^{-3}$ J. The neodymium glass version CISE II, with a flash tube voltage of 1·2–1·5 kV, had an output of 0·2 J; with an achromatic lens of 32 mm focal length at a magnification of $\times 4$, an energy density of $10^8$ W/cm$^2$ was achieved.

The production model LMA 1 of VEB Carl Zeiss is equipped with the ZFL 1000 Laser (Fig. 3(b)). The various supply and servicing units and the cooling system for continuous use are mounted in a movable cabinet. Inside the cabinet are the blower for cooling (at the back, on top) and three laser units [capacitor bank (Fig. 5, part 4), charging unit (part 3) and operating unit (part 1)] and the combined charging and operating unit for the excitation electrodes on the LM microscope (part 2). The chassis with the capacitor bank of the laser, running on rollers, is inserted in the bottom section of the console behind a removable panel. It contains four paper capacitors of 127 μF, the inductances, an Ignitron safety switch and the electrical components for the step control (127 – 254 – 386 – 508 μF), for the Ignitron and the pumping flash tube ignition. On the front of the console, behind a door with a lock, are four switches for initial control of the pump energy and of the discharge characteristics through appropriate settings of the $C$ and $L$ values.

Above the chassis of the laser capacitor bank is that of the laser charging unit (Fig. 5, part 3). It contains the rectifier, the high-voltage transformer for charging the capacitors, the charging controls, the control amplifier for capacitor voltage stability and several other components. The laser operating unit (Fig. 5, part 1) contains the voltage setting control and voltmeter. The selector switch for the laser has four setting positions for different release modes.

The resonator head (Fig. 5, part 5) can by means of a rapid change mount be fitted either directly to the microscope LM (see § 2.3) or to the optical adaptor of the LMA 1 outfit (see § 2.3.2). A Q-switching device for special resonators can be connected to the resonator head through another rapid change mount, which is in a dovetail ring form. Included in the resonator head are the laser resonator itself, and the pumping source, a Pressler straight flash tube G 455 (Table 1); they are placed in the positions of the focal lines of a reflector in the shape of an elliptical cylinder, and can easily be changed. The resonator is a cylindrical neo-

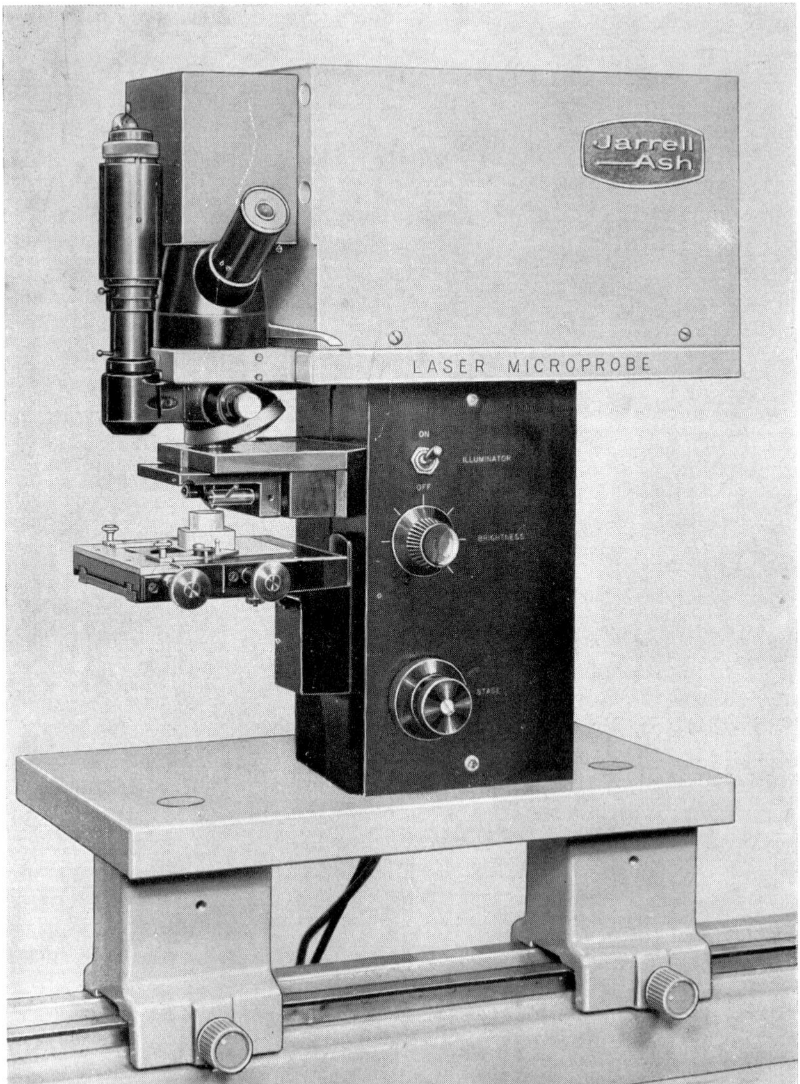


FIG. 4. Laser microprobe of Jarrell–Ash Co., Waltham, Mass., USA. (*a*) Laser head and microscope

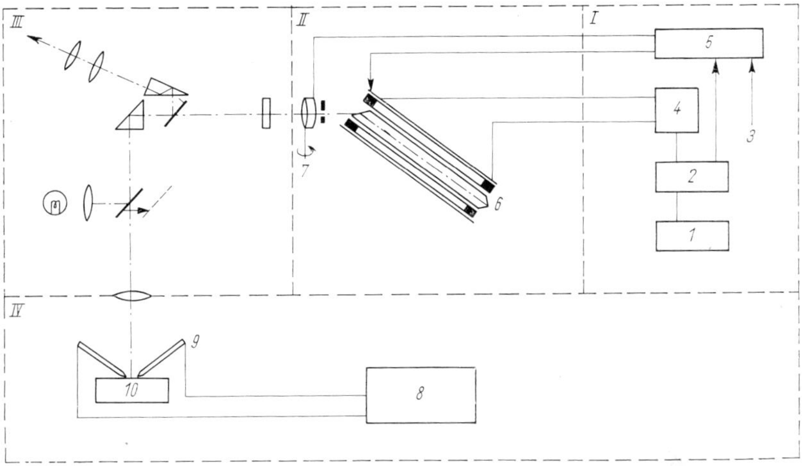


FIG. 4.

(*b*) Diagram of system:

I Laser supply unit II Resonator head III Microscope for reflected light
IV Auxiliary electrodes with supply unit:
1 Mains supply unit; 2 High frequency regulator; 3 Trigger unit; 4 Capacitance and inductance; 5 Main control; 6 Resonator, pumping flash tube and reflector; 7 Active switch; 8 Supply unit for spark gap; 9 Electrodes; 10 Sample.

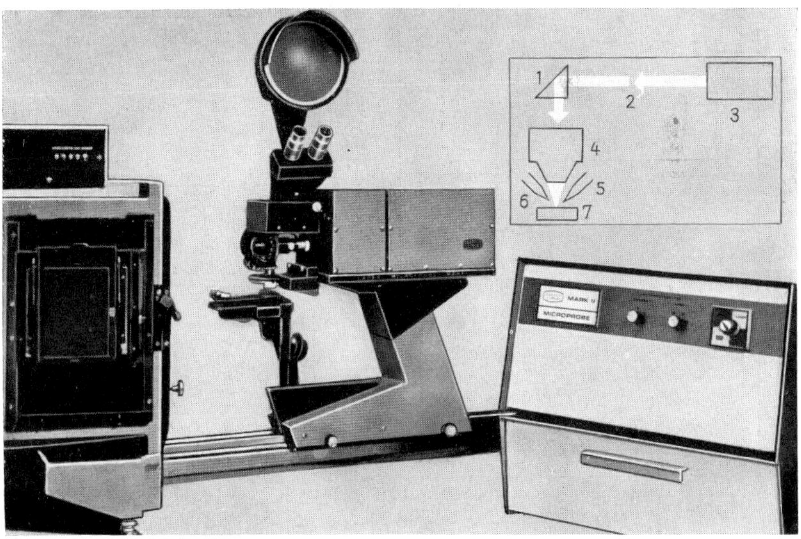


(*c*) Jarrell–Ash laser microprobe Mark II (1966) on optical bench of Czerny–Turner spectrograph.

1 Prism; 2 Laser radiation; 3 Resonator head; 4 Microscope; 5 Electrodes; 6 Microplasma; 7 Sample.

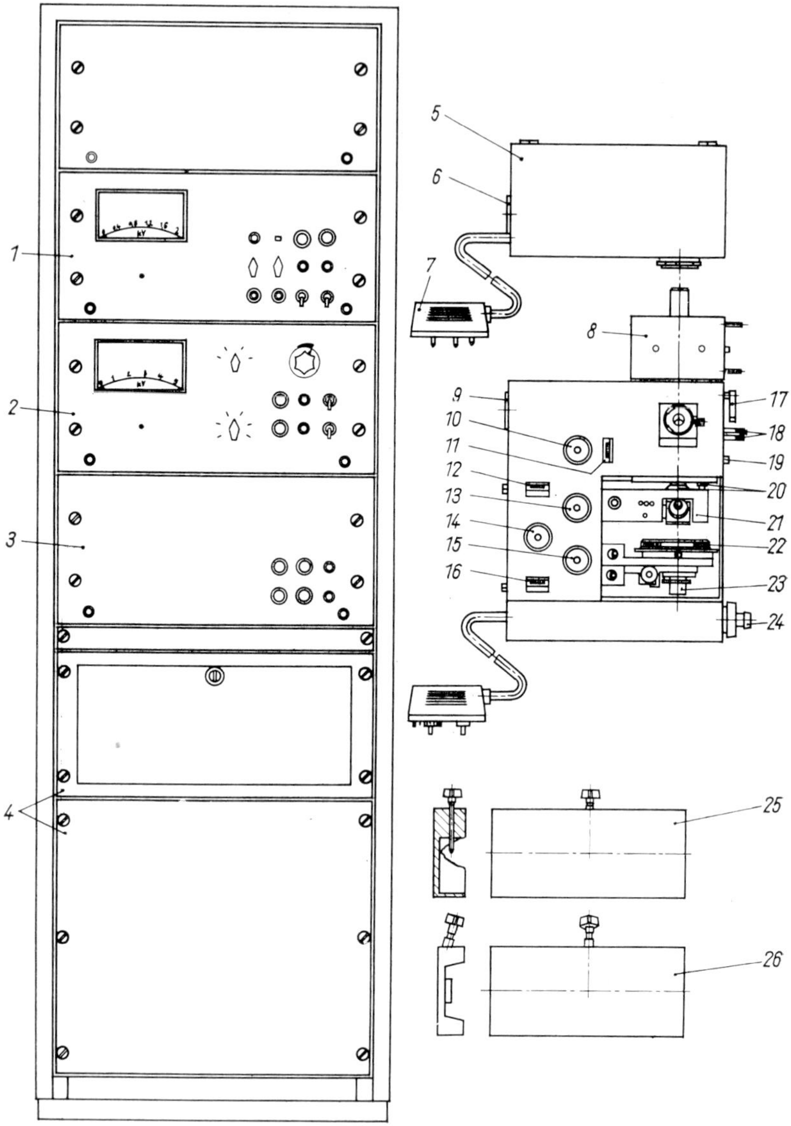
1
2
3
4
5
6
7
8
9
10
11
12
13
14
15
16
17
18
19
20
21
22
23
24
25
26

dymium glass rod (9·5 mm diameter and about 110 mm total length) with one end worked as a totally reflecting 90° roof prism and surrounded by a protecting outer glass tube. The plane exit surface of the resonator with a partially reflecting coating can be adjusted by autocollimation to be accurately normal to the optical axis of the microscope; the resonator can be tilted within its mount about two axes in the plane end surface.

Ruby resonators of 9·5 mm diameter and 105 to 110 mm length with coated plane end surfaces (without use of reflecting prism) can also be used in the laser head of the LMA 1 outfit. The standard Zeiss laser for this outfit includes a spare neodymium glass resonator. The unit has a maximum consumption of about 5·5 A, a maximum charge energy of 1000 J, a maximum output energy of 1 J and a pulse repetition rate of 3 $min^{-1}$. The flash tube voltage can be continuously varied between 1 and 2 kV; the four alternative capacitance levels available and the choice of two inductances have already been mentioned. The supply and operating unit for the excitation electrodes on the microscope is discussed in § 2.4.

The first commercial laser microprobe of the Jarrell–Ash Company, Waltham, Mass. USA[54–56] is an instrument of a different type in which the laser is used with controlled feedback. It is fitted with the Q-switched ruby laser made by Trion (now Lear Siegler Inc., Ann Arbor, Michigan). The essential technical features of this analytical outfit have been described in 1967 by Rasberry, Scribner and Margoshes.[104,105] The 4 kV mains supply unit has a capacitance of 125 μF and a charging period of 30 s. The end surfaces of the ruby resonator (57 mm length and 5 mm diameter) are set at the Brewster angle to reduce the reflection losses. The

---

FIG. 5. Diagram of VEB Carl Zeiss, Jena, laser micro-spectrum analyser LMA 1, comprising mains supply console, microscope with optical adaptor and resonator head.

1 Chassis for laser ZFL 1000; 2 Chassis for cross-excitation electrodes; 3 Charging chassis; 4 Capacitor bank chassis; 5 Resonator head; 6 Circular dovetail connector for Q-switch attachment; 7 Connector plug for resonator head supply; 8 Optical adaptor with diaphragm wheel and adjusting screws; 9 Rapid change connector for photomicrography attachment; 10 Selector knob for autocollimation alignment, visual observation and photomicrography; 11 Illumination field stop for reflected light; 12 Filter slide and aperture stop for reflected light; 13 Setting knob for electrode holder; 14 Selector knob for transmitted light, electrode projection and reflected light; 15 Setting knob for microscope stage; 16 Filter slide and illumination field stop for transmitted light; 17 Selector lever for free passage of laser radiation or visual observation, photomicrography and resonator alignment; 18 Analyser, whole- and quarter-wave plate; 19 Selector slide for two objectives; 20 Two objectives in selector slide; 21 Electrode holder with setting controls; 22 Mechanical microscope stage; 23 Microscope condenser for transmitted light; 24 Coarse and fine focusing of microscope objectives; 25 Adaptor for Q 24 quartz prism spectrograph and for three-prism spectrograph; 26 Adaptor for plane grating spectrograph PGS 2.

resonator is housed in a cylindrical reflector and pumped by a helical flash tube. A prism rotating at a speed of 17000 r.p.m. produces a pulse power of about 4 MW. The reported pulse duration is 35 ns. In an alternative mode of operation 2 to 6 main spikes with an output of about 1 MW occur in intervals of 100 to 200 ns. An energy density of $10^{14}$ W/cm$^2$ has been determined at the focus of the microscope objective $\times 20/0{\cdot}25$ with a 13 mm working distance.[103] According to Rasberry *et al.* the energy density in the parallel laser beam is 3 to 4 $\times$ $10^7$ W/cm$^2$. The laser can only be triggered when the flash tube capacitors are charged to a preset voltage and when the air around the ruby rod has reached a preselected temperature of 0 to 20 °C $\pm$ 0·5 °C above room temperature. The laser beam emerges from the resonator through a set of sapphire plates.

The circuit parameters of the auxiliary electrode gap are: 1·5 to 2 kV, 0 to 155 μH, 0 to 10 Ω, 5 to 10 μF.

The laser head and the metallurgical microscope form a heavy unit which is best mounted on the optical bar of the grating spectrograph (Fig. 4). The power supply and capacitor banks and the trigger system are housed together in a separate electronic console.

The laser microprobe Mark II, Model 45-604 (April 1966) (Fig. 4c), differs from the first Jarrell–Ash microprobe (May 1963) in that two straight flash tubes, fitted in two elliptical reflectors, are used in the later version. The makers encountered difficulties in cooling the rubies[3] and therefore now only use neodymium glass resonators of 150 mm length giving a beam of 20 minutes of arc divergence.

The laser power unit, for a mains supply of 115 V a.c., 50 to 60 Hz, is table-mounted and has dimensions of 485 × 510 × 230 mm. An alternative version mounted on rollers for ease of handling can be supplied to order. The capacitor bank contains three General Electric capacitors, 25 μF/4 kV; the flash tube voltage can be set to 2·2, 2·3, 2·5 and 2·9 kV. The supply unit takes a load of 350 W. The optical resonator delivers an energy from 0·1 to 1·0 J; the pulse power is about 0·4 MW, with a pulse duration of 1 to 2 μs and a repetition rate of several pulses per minute.

The optical system of the Lear Siegler laser LM-2, adopted by Jarrell–Ash, is contained in a sealed unit; the user cannot change resonators or flash tubes himself. There is no external air or water cooling.

The advantages of a pulse counter are almost lost for routine work because altogether thirty screws have to be removed to take off the two sheet metal covers of the laser head.

The laser of the microprobe model MSL 2 of the USSR, LOMO, Leningrad, October 1967 (Fig. 6)[72] has a flash tube of 200 mm length with an electrode separation of 40 mm, taking a load of 400 J. The emission wavelength is 1·060 μm; the pulse duration is 500, 625 or 750 μs and can, by means of a prism rotating with a speed of 24000 r.p.m., be reduced to 30 ns. The KGSS/7 glass resonator of 6 mm diameter yields with Q-switching 2 MW and without Q-switching a maximum pulse energy of

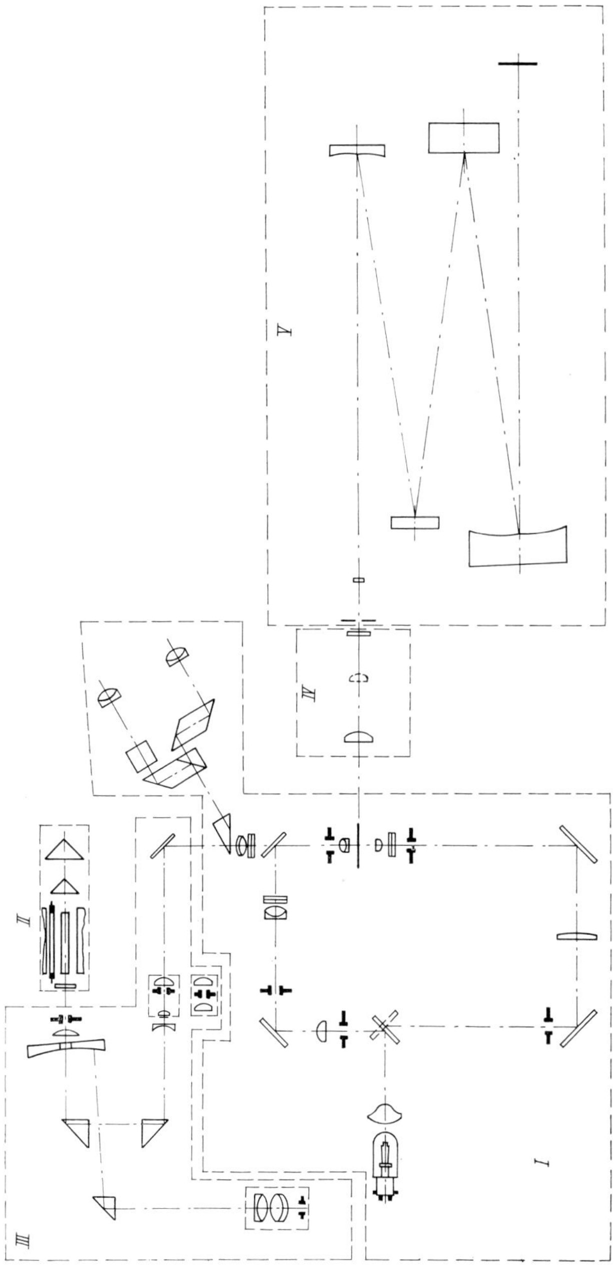


FIG. 6. Optical diagram of U.S.S.R. laser outfit MSL-2.

I Microscope. II Laser head with two prisms. III Adjusting system for laser and microscope optic. IV Condensers. V Grating spectrograph.

0·4 J. The maximum electrode voltage of the flash tube is 2 kV. Resonator and flash tube are cooled by a double water circuit. The capacitance of the laser discharge circuit is 200, 300 or 400 μF. The resonator has on the output side an external reflector. The laser optic is aligned by autocollimation. The resonator head is rigidly attached to the microscope stand. On the front of the microscope is a protective sheet metal cover; if it is closed no adjustment, such as focusing or stand alignment, can be carried out.

Since 1967 Japanese firms too have been developing apparatus for spot analysis.[6,60] Shimadzu Seisakusho have designed an instrument with a 5 J laser. The liquid-cooled ruby of 100 mm length and 8 mm diameter is excited by six xenon flash tubes. A crypto-cyanine absorption cell is used for passive Q-switching.

A Jeol model has a high performance ruby laser JLR-O2A with passive Q-switching, an output of 70 MW and pulse duration of 15 ns. A laser output of 20 MW and energy of 1·5 J (for resonator dimensions of 6·35 mm diameter and 76·2 mm length) are also reported by other sources. The pulse repetition rate is 12 $min^{-1}$. Katsuno *et al.*[60] describe a focusing lens system with magnification ×10. No information on the dimensions of the craters obtained is given anywhere, even in the Jeol publications. The listed examples of applications, such as determination of Cu, Ni, Mn and V in steels and of Cu, Si and Fe in aluminium alloys, indicate that this second instrument developed in Japan is intended for the routine analysis of industrial products (macro-analysis, evaluation of alloy constituents).

Q-switched lasers of lower pulse power have been used for spectrum analysis. In 1965 Laqua, in Jena, reported how with a so-called non-linear filter one can obtain a pulse power of 200 kW, and with a rotating interferometer mirror (3000 r.p.m., duration of 'opening' 30 ms, hence not really a 'giant pulse') control the crater dimensions at the same energy output of 0·3 J. With the same solid-state laser without Q-switching a pulse power of 2 kW has been measured under the following operating conditions: elliptical reflector, input energy between 320 and 420 J, ruby threshold value about 200 J—half of this value for neodymium glass—focusing with Vickers achromat of magnification ×20, numerical aperture 0·25 and focal length 8 mm. The pulse repetition rate of one pulse per second in 1966 was stepped up to 12 pulses per second by means of a special water cooling system ('swimming pool'). This is of special importance for laser macro-spectrochemical analysis.

According to Laqua output energies between 0·6 and 2 J have been obtained with a ruby laser at the Institute for Spectrochemistry, Dortmund. In analytical work the laser radiation has been attenuated by filters to obtain output energies of the order of 20 mJ at the sample surface.

The Q-switched pulsed laser used in 1963 by Runge, Minck and Bryan of the Engineering and Research Staff of the Ford Motor Company, Dearborn, Michigan, for emission macro-spectroscopy had the following specification:[113] power output between $3 \times 10^6$ and $5 \times 10^6$ W, pulse duration 15 to 20 ns, output energy between 50 and 75 mJ, energy density at the focus of a high power lens $10^{13}$ W/cm$^2$.

Archbold, Harper and Hughes[7] also used a Q-switched ruby laser in their examination of the microplasma generated by long wavelength, red radiation. The resonator had a length of 70 mm and a diameter of 9·5 mm. It was pumped in a spherical reflector by a collinear xenon flash tube. One end of the ruby rod was worked into a totally reflecting 90° prism, the other end was flat and uncoated. The laser used at the Light Division of the National Physical Laboratory, Teddington, Middlesex had a power output of 1 MW.

Development and production of lasers have progressed rapidly and a great variety of systems covering wide energy ranges have been made available. An appreciable number of models have been short lived. It is difficult—and indeed not necessary—to give a precise up-to-date account of the present state. Only a short survey of some small lasers useful for emission spectroscopy is made in the following pages, partly to illustrate the course of development, without any claim to being complete and up-to-date in every detail.

Some of the lasers with a pulse output energy up to 1 J have no cooling of the resonator housing, hence a low pulse repetition rate. The 103-Mini-Laser model has an output of 0·005 J with a pulse duration of 500 μs and a repetition rate of 1 pulse per minute. The TRG Inc., Syosset, N.Y., joined the laser head to the research microscope 'Ortholux', forming, together with a photomicrography attachment, the Biolaser Model 503. It has a mains supply unit which provides input energy for the flash tube of 200 J. The pulse repetition rate can through cooling be increased to $\frac{1}{2}$ s$^{-1}$.[127]

The Laboratory Laser Model 120 of the Optics Technology Inc., Palo Alto, Cal. (Fig. 7a) has a ruby rod of 6·35 mm diameter and 76·2 mm length mounted in an elliptical reflector and emitting pulses of 0·1 J energy. The supply unit provides an input energy of 400 J. A liquid nitrogen cryostat is available as an accessory.[94]

The Model 190 Microscope Laser made by the same firm (Fig. 7(b)[94]) has a similar performance. Peppers[97] has described in detail an outfit consisting of this laser together with a microscope. He belives that even a 30 mJ laser pulse can damage cemented optical components. His view was taken into account in designing this American instrument; the ruby resonator of the same dimensions as that of the Model 120 emits a maximum output energy of 0·5 J with a pulse duration of 200 μs. The laser head of 432 mm length and 76·2 mm diameter is air cooled. Two straight flash tubes are mounted in direct contact with the resonator. The repetition rate is 12 pulses per minute. An iris diaphragm is incorporated in the resonator housing; this allows adaptation of the set-up for use with different objectives and, with a diameter of about 1 mm, only lets pass

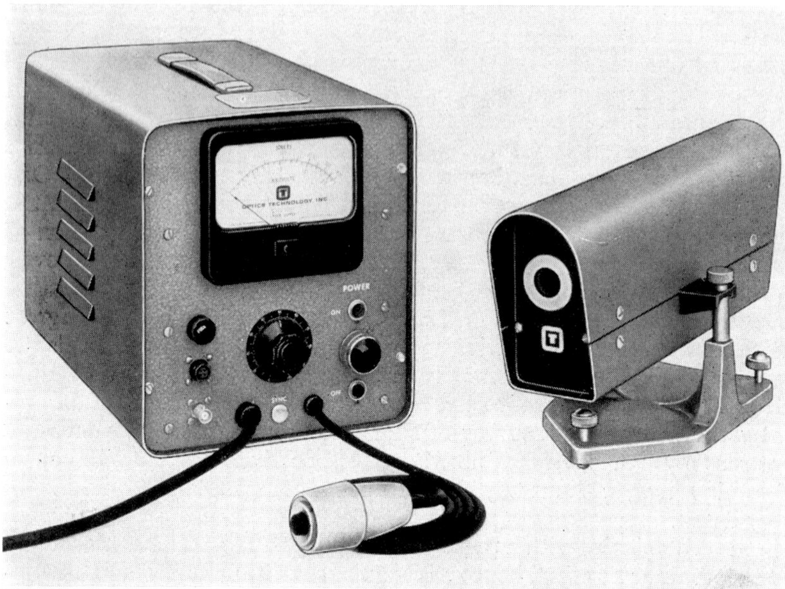

FIG. 7. Two solid-state lasers of the Optics Technology Inc., Palo Alto, Cal., U.S.A. (*a*) Laboratory laser model 120 (left, mains supply unit; right, laser head).

0·4 per cent of the energy emitted by the ruby. Fig. 7(c) shows a diagram of the optical path. A negative lens is placed in the back focal plane of the objective, between the latter and an ×10 eyepiece; the lens has a 15-layer dielectric coating of 50 μm diameter acting as a divergence reducing diaphragm. This special coating transmits radiation below the wavelength 650 nm almost completely, but only lets 0·3 per cent of the laser emission pass. The area of action on the sample surface is thereby limited, and at the same time the test detail to be investigated is exactly outlined. Of the energy emitted by the resonator only a fraction of 30 μJ remains after it has passed through the objective. The diameter of the crater produced by an objective system ×40/0·65 in various metals and biological samples is about 1 to 2 μm.[97] The quantity of material required for micro-spectrochemical analysis with a conventional spectrograph can probably only be obtained with difficulty and in particular cases with this outfit. Therefore the design allows removal of the diaphragm forming part of the standard outfit. Without it crater diameters of up to 40 μm can be obtained as a result of the divergence of the laser beam.

The excitation energy of 625 J for the straight flash tube of the Laser Model 190 is taken from a fairly small mains supply unit. In addition to the ruby resonator a neodymium glass rod is supplied. An energy density

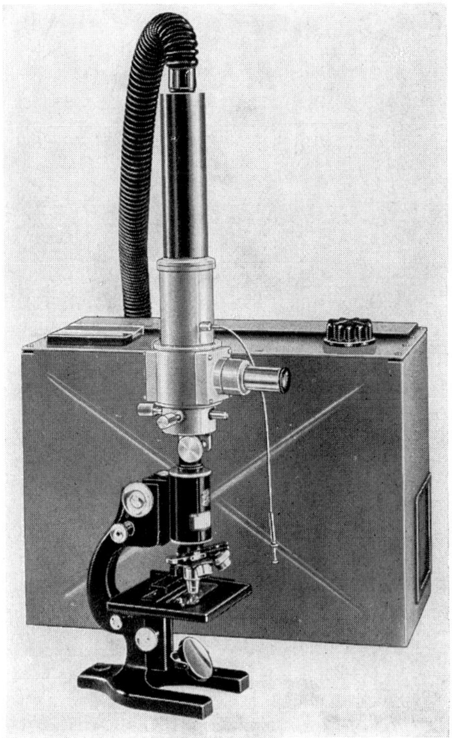

FIG. 7b

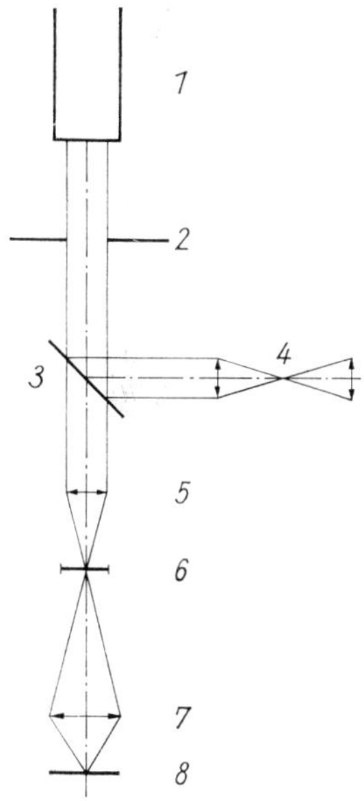


FIG. 7c

(*b*) Model 190 Microscope Laser with mains supply unit.
(*c*) Optical diagram of Model 190 Microscope Laser.
1 Resonator; 2 Iris diaphragm; 3 Dichroic mirror; 4 Observation telescope; 5 ×10 eyepiece; 6 Negative lens with dielectric coating serving as divergence reducing diaphragm; 7 Microscope objective; 8 Sample.

of 300 J/mm$^2$ has been measured at the focal spot of the microscope laser of 1 μm diameter. The same energy density is obtained with other similar designs used mainly for biological investigations.

Bessis[14] at the Centre National de Transfusion Sanguine, Paris, irradiated blood corpuscles with a CSF laser that had a ruby rod of 3 mm diameter and 40 mm length. The divergence of the laser beam had been found to be $5 \times 10^{-3}$ radian. In the set-up the resonator rod was mounted in front of a ×6 objective; the effective laser cross-section was reduced by the eyepiece diaphragm to two thirds of its full size. With an objective ×100/1·25 a focal spot of 2·5 μm diameter was produced. The flash tube of the CSF laser requires a voltage of 2·5 kV; the capacitance of its capacitors is given as 160 μF.[5]

Several firms throughout the world produce ruby lasers in the 1 J range. The standard outfits are usually mounted on a simple optical bench. This is not a practical arrangement for use of the laser with a microscope.

## 2.3. MICROSCOPES

The first commercially produced outfits for laser micro-spectrochemical analysis included a reflection microscope (a microscope working with light reflected by the sample) (Jarrell–Ash Co.) or a standard research microscope modified for work with reflected light (VEB Carl Zeiss, Jena).

Microscopes of this type can be designed with inverted stands (Le-Chatelier type; for example Zeiss Neophot and Zeiss Epityp) or in the conventional manner, with the microscope tube above the stage and with illumination of the sample also from above.

Microscopes of the Le-Chatelier type with their firmly fixed, cemented components of imaging and deviating optics are not very suitable for use with a laser. The comparatively simpler optic of metallurgical microscopes with illumination for sample reflection (for example the Zeiss Epignost) or of the medium research microscopes, nowadays usually also equipped for work with transmitted or reflected light (for example the Nf Microscope of VEB Carl Zeiss), can easily be adapted for the fundamental requirement of a laser microscope: namely, to avoid the monochromatic long wavelength radiation passing between the upper end of the microscope tube and the entrance pupil of the objective through optical components with easily melted cement layers.

There are two alternatives for the construction of the vertical illuminator: the optical axis of the condenser for work with reflected light can be at right angles to the microscope tube axis (Zeiss microscopes) or parallel to it (Jarrell-Ash microscope).

A microscope stand cannot as a rule be mounted directly on an optical bench. An adequate laboratory outfit for micro-spectrochemical analysis can be assembled by bridging the optical bench with an adaptor plate made for the purpose; on it the microscope in its standard form can be positioned in the lightbeam in front of the spectrograph slit, between the electrode projecting unit and the first condenser lens. The assembly can be made more rigid by taking off the microscope stand and securing the body on a base plate which carries a support arm for the laser head. The base plate itself rests on an adaptor block shaped to the profile of the optical bench (Fig. 8a and 8b). Such a mount, constructed in 1964 by VEB Carl Zeiss at the suggestion of the authors, fitted the optical bench with a triangular cross-section of the Zeiss prism spectrograph Q 24 and the three-prism glass spectrograph. The base plate could be turned through small angles for aligning the electrodes. The adaptor body could be clamped in position and had an electrical terminal for earthing. An alternative adaptor body was available that fitted on the square cross-section optical bar of the 2 m Ebert plane grating spectrograph.

The laser microprobe designed by Brech and Schuch[55] has a base plate that rests on two carriages mounted on the optical bench of the Jarrell-Ash 1·5 m Wadsworth grating spectograph.

In 1964–1965 VEB Carl Zeiss adapted their Nf stand for laser microanalysis. It was fitted with various experimental holders for the auxiliary electrodes and mounted on an adjustable baseplate as described. The Nf microscope is the research model of the N Series of VEB Carl Zeiss that

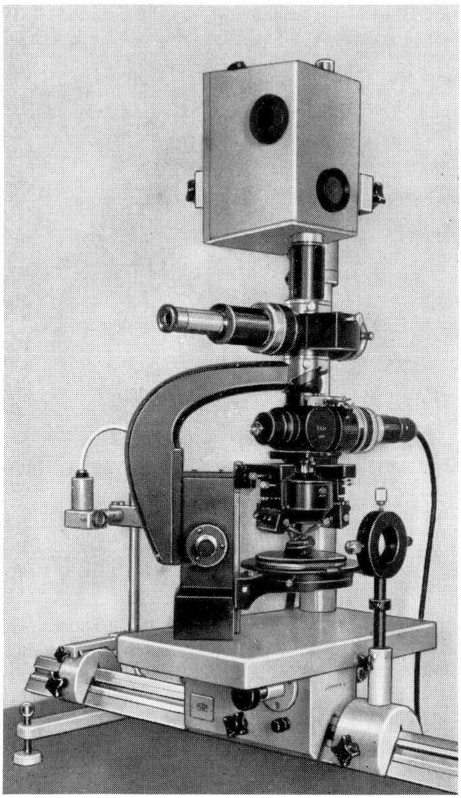

FIG. 8(*a*). Modified Nf microscope with electrode holder and with Zeiss solid-state laser head ZFL 750 on adjustable base plate (December 1964). Also mounted on the optical bench are an electrode projecting unit and a quartz condenser lens (80 mm focal length).

was developed between the L Series and the X Series. The stand has been retained in the form of the L Series. Important new features are a device for rapid changing of the sample stage, coaxial coarse and fine drives, ball bearing for fine motion carriage, removable condenser tray, condenser drive body and stage base. Model W of the latter takes two single motion stages and one co-ordinate stage type K 1 with a movement range of 76 × 50 mm. The authors' experience showed both stage types to be suitable for micro-spectroscopy with the modified Nf microscope. The authors used in particular the mechanical stage model H 5 because its dimensions allow imaging with condensers in their normal position and mounting.

The maximum acceptable sample height was increased by fitting a 50 mm metal block between the microscope drive body and the base plate. The length of the pinion rack had to be increased at the same time.

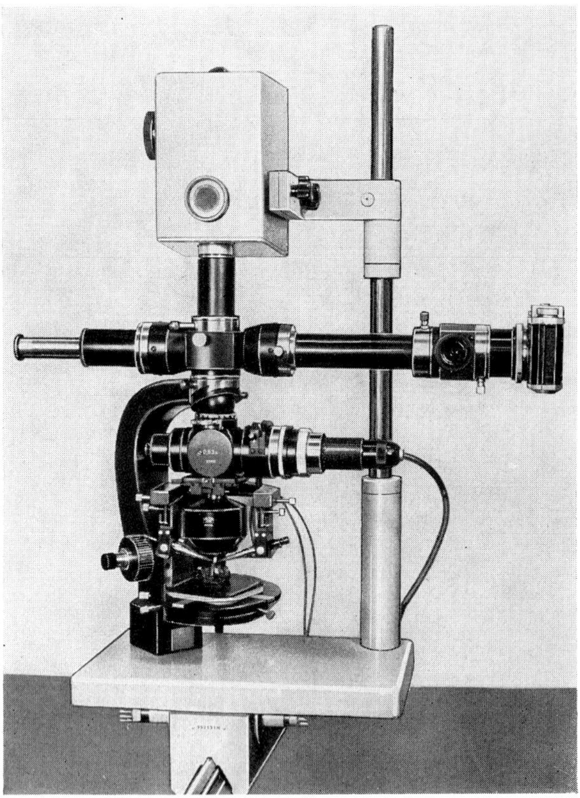

FIG. 8(*b*). The first Zeiss laser micro-spectrum analyser with photomicrography unit (24 × 36 mm).

In the Nf microscope the electrode holders of the spark gap were clamped to the rail of the fine motion drive that acts on the sample stage; hence focusing the sample brought the electrodes out of alignment. The guide rail had therefore to be shortened at the top and an additional short rail fitted which was independent of the fine motion drive.

In their first investigations the authors used a very simple electrode holder, which was basically a sample base W. To the base were added two metal holders which could be rotated and clamped in any position; these holders took plastic rods with bore holes in which the carbon electrodes were fixed. Other holders are described in § 2.4. Attached to the lower end of the microscope tube mount is a rapidly interchangeable reflected light condenser, model 'Pol', with polarizing filters, and attached to the upper end of the mount a modified MF body unit with a three-position selector. With this unit, containing two deviating prisms, in the middle position the parallel laser beam passes unaltered through the microscope. The two other positions serve for photomicrography and for visual sample observation through a monocular viewing tube, with or without an angle piece, with a magnification factor of $\times 1{\cdot}6$, giving, together with the $\times 0{\cdot}63$ condenser for reflected light, a factor of 1. The selector slide carriage for the objectives used (see § 2.3.1) are inserted into the underside of the condenser. A light-tight protecting tube is fitted between the main body and the resonator holder of the laser head.

The different nosepieces of the series N microscopes take objectives of the basic outfit, suitable for a 160 mm tube length, namely: achromats, apochromats, flat-field achromats for work with light transmitted through the sample. If a Nf microscope is mounted on an adjustable base plate as described above, polished thin sample slices and other transparent test pieces can also be spectrochemically analysed: an illumination system is built into the stand of the Nf microscope, with an adjustable front-surface deviating mirror, ground glass screen and iris diaphragm. A 30° angle piece with tube factor 1 is used for viewing transmitting objects.

Trucco[128] constructed an outfit for spot analysis with two Reichert microscopes—Neopan and Metatest—using two achromats of magnifications $\times 4$ and $\times 32$.

The prototype of VEB Carl Zeiss, Jena, with the Nf microscope for reflected light has been replaced by a microscope of a new design, mounted on an optical bench (Fig. 9) which has several advantages. This microscope model LM of the laser micro-spectrum analyser LMA 1 consists of a microscope stand transversely adjustable, with a built-in illumination system for various methods of observations by reflected and transmitted light, chosen by a selector switch. A number of components can be attached to this stand, such as a condenser mount for illumination by transmission, a support for a mechanical sample stage with wide range height adjustment, an adjustable electrode holder mount, a visual observation tube, a photomicrography attachment, an optical adaptor and the laser resonator head.

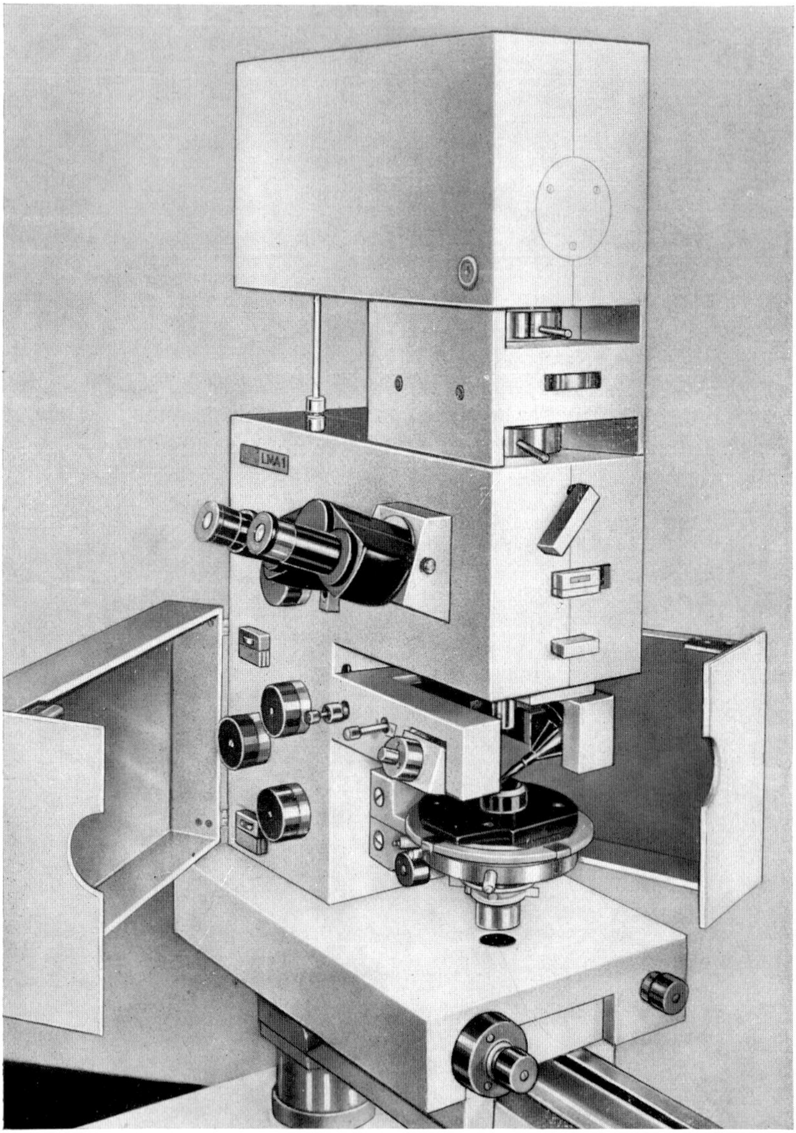


FIG. 9. LMA 1 laser microscope for micro-spectrochemical analysis with resonator head ZFL 1000 and optical adaptor. VEB Carl Zeiss, Jena.

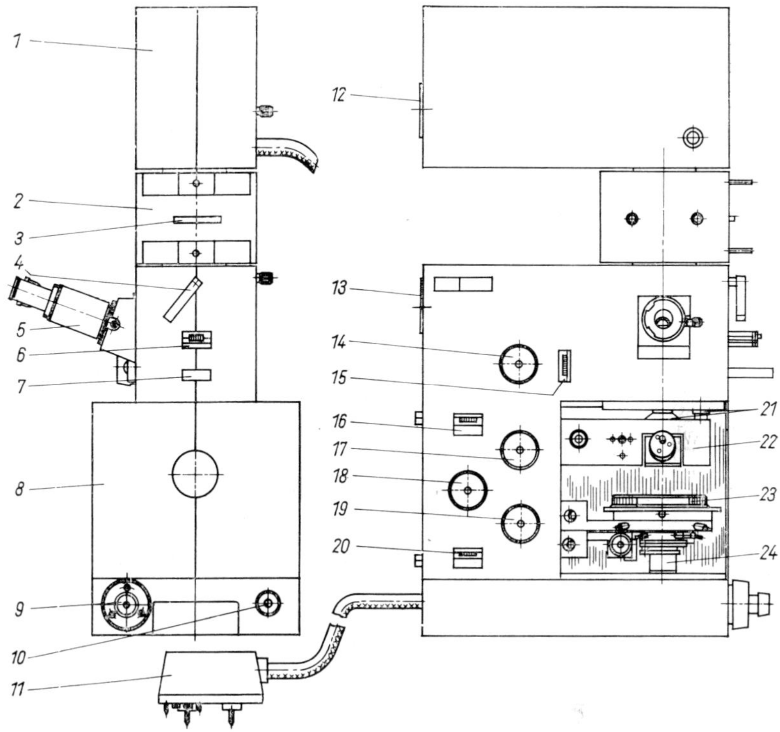


FIG. 10. Diagram (front and side view) of microscope of LMA 1 outfit with resonator head.

1 Resonator head; 2 Optical adaptor; 3 Diaphragm wheel; 4 Selector lever for free passage of laser radiation or visual observation, photomicrography and resonator alignment; 5 Binocular eyepiece tube; 6 Analyser, whole- and quarter-wave plate; 7 Selector slide for two objectives; 8 Safety doors; 9 Coarse and fine focusing of microscope objectives; 10 Knob for turning the microscope mount; 11 Connector plug for microscope supply; 12 Connection for Q-switch attachment; 13 Connection for photomicrography unit; 14 Selector knob for autocollimation alignment, visual observation and photomicrography; 15 Illumination field stop for reflected light; 16 Filter slide and aperture stop for reflected light; 17 Setting knob for electrode holder; 18 Selector knob for transmitted light, electrode projection and reflected light; 19 Setting knob for microscope stage; 20 Filter slide and illumination field stop for transmitted light; 21 Two objectives in selector slide; 22 Electrode holder with setting controls; 23 Mechanical microscope stage; 24 Microscope condenser for transmitted light.

The complete microscope LM rests on an interchangeable base with an adaptor for the triangular optical bench of the quartz spectrograph Q 24 and the three-prism spectrograph, or for the square optical bench of the Ebert plane grating spectrograph PGS 2. The base is secured in position on the optical bench by a clamping screw (Fig. 5, parts 25 and 26).

As a safety measure against risks from the high voltage of 1·6 kV or more across the charged electrodes the middle section of the stand is screened off by two doors with glass windows which are kept open in ordinary microscope work. The voltage is applied to the electrodes only when these doors are closed and when the selector switch on the side of the stand points backwards; this switch sets the outfit alternatively for free passage of the laser radiation or visual observation, photomicrography and resonator adjustment (Fig. 10, part 4).

The diagram Fig. 10 of the front and side views of the laser microscope shows the coaxial coarse and fine focusing knobs, acting on the objective slide (7), which are positioned at the bottom of the right side panel and, on the front, the binocular eyepiece tubes (5), the knurled control disks (15 and 16) for the aperture and field stops for reflected light and that for the field stop for transmitted light (20). Other important controls are the various setting knobs, namely the three-position selector (14) for, from

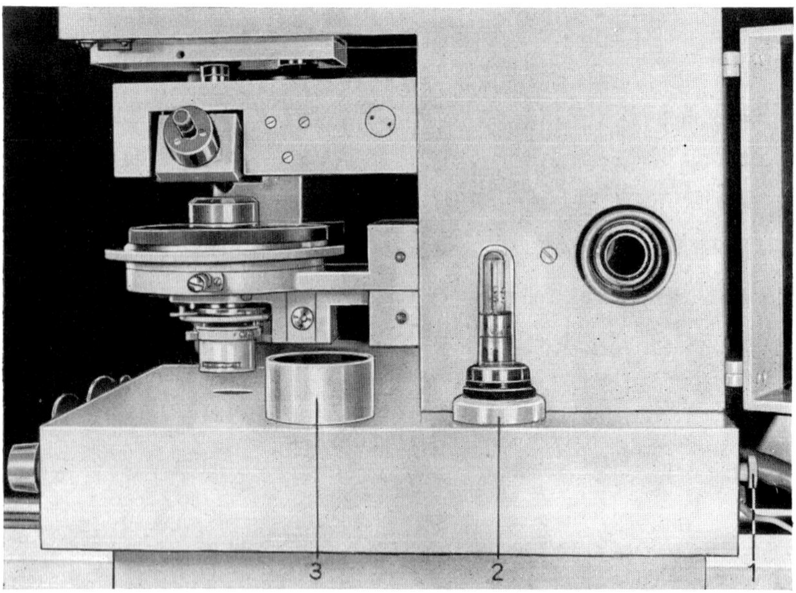


FIG. 11. Back view of microscope stand LM.

1 Earthing connection; 2 Lamp holder with coiled ribbon filament lamp; 3 Lamp holder funnel.

left to right, laser resonator adjustment, visual observation, photomicrography; the three-position selector (18) for, from left to right, work with transmitted light, electrode projection, work with reflected light; the knob (17) for height adjustment of the electrode holder; the knob (19) for coarse setting of the sample stage (23) and transmission condenser (24). The two slides (16) and (20) take filters that may be required for observation in reflected or transmitted light.

The electrodes are clamped in position in the holder which is provided with three adjustment controls. Within the fork-shaped electrode holder the reflecting objective ×40/0·5 is visible in Fig. 9, and also the second objective ×4/0·05 which serves for accurate setting of the electrode separation appropriate for exciting a vaporized sample quantity and for various preliminary and general observations. The second objective is fitted in an adjustable holder and is interchangeable with the lens system ×16/0·2 provided, like the reflecting objective, for laser irradiation, photomicrography and visual observation.

Above the objective selector slide (Fig. 10, part 7) are two further slides (6); the upper one brings the analyser into the beam, at the same time introducing the polarizer into the illumination beam for work with reflected light (the polarizer for transmitted light is permanently incorporated in the microscope stand). With the knurled disc in this slide the analyser can be rotated through 90°. The lower slide (6) has three setting positions for a clear aperture (uninterrupted light passage) and for a quarter-wave plate and a whole-wave plate respectively.

Fig. 10 shows how the optical adaptor, discussed in § 2.3.2, and the laser resonator head are connected to the microscope LM by rapid change mounts.

The illumination comes from a 6 V, 15 W coiled ribbon filament lamp which can be unscrewed and changed by means of the ring (2), Fig. 11. In use this ring is covered by the funnel (3) which is screwed on to it. Fig. 11 also shows, next to the entering high voltage cable, the clamp screw (1) connecting the microscope housing to the earthing of the laser mains supply unit.

As can be seen from this description and the optical diagram of Fig. 12,

TABLE 3. Specification of the optical systems of the Zeiss microscope LM

| *Non-immersion types* | *Objectives Lateral magnification* | *Numerical aperture* | *Eyepieces Angular magnification ×8, Field size number (16)* | *Eyepieces Angular magnification ×12·5, Field size number (16)* |
|---|---|---|---|---|
| Flat-field achromat | × 4 | 0·05 | 32 | 50 |
| | ×16 | 0·20 | 125 | 200 |
| Reflecting objective | ×40 | 0·50 | 320 | 500 |

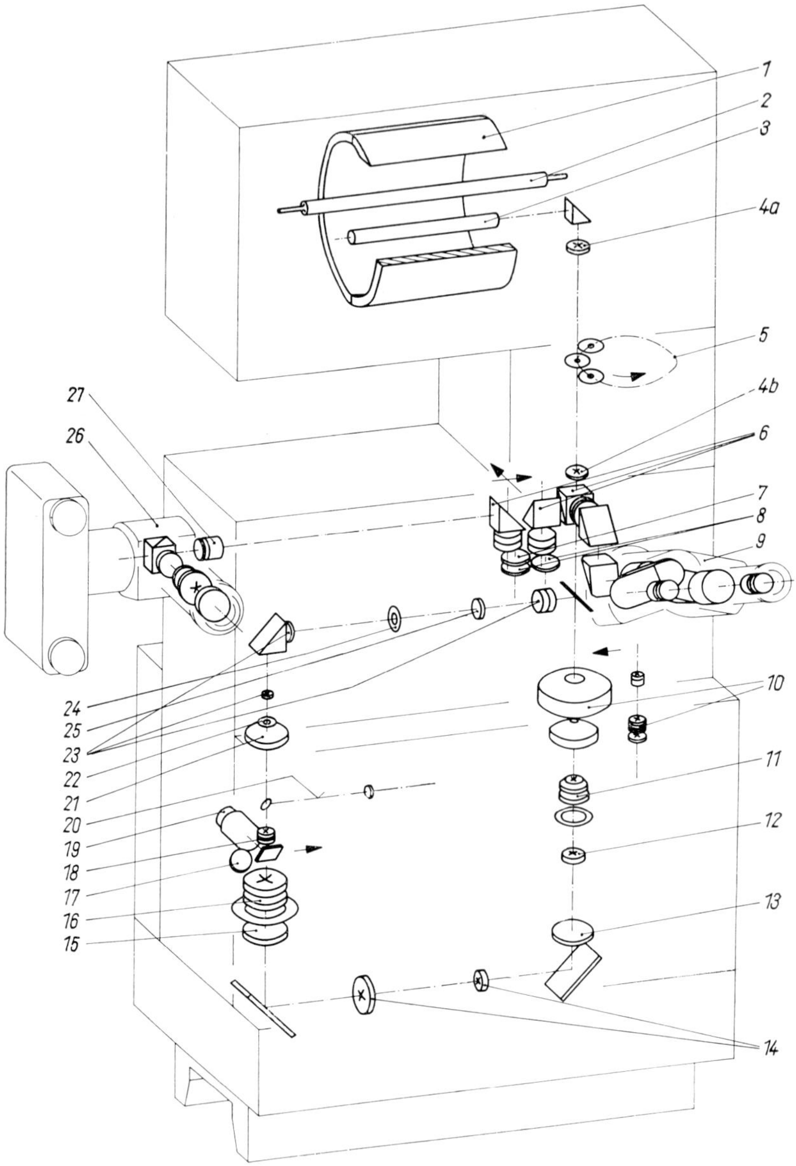
1
2
3
4a
5
4b
6
7
8
9
10
11
12
13
14
27
26
24
25
23
22
21
20
19
18
17
16
15

in the Zeiss microscope model LM the laser radiation falls vertically from above on to the sample which can be visually observed by the bright field and polarisation method either in reflected or in transmitted light. The condenser for the transmitted light can be removed to increase the available sample space to a height of 70 mm.

The specification of the available optical systems is given in Table 3.

Since 1966 all standard instruments have been fitted with a removable substage (tray) for the wide-angle-lens condenser 0·6 for transmitted light. The substage comprises a clamping attachment mount, focusing drive and condenser tray. Fig. 13 shows the case with the various microscope components and accessories. Some of the parts are illustrated separately in Fig. 14: the mechanical stage with its support and condenser tray, both for clamping on to the microscope column; the condenser and the fitting tool. Samples of dimensions up to 100 × 100 × 75 mm and of maximum weight of 2 kg can be accommodated.

The 1964 microprobe of Jarrell-Ash Co. is a compact instrument. The metallurgical microscope and the Q-switched laser are housed in a single unit. Sample examination by transmitted light is not possible. In 1966 the stand was modified. It now is Z-shaped and fits on the standardized optical benches of the Jarrell-Ash spectrographs. Furthermore a triple eyepiece unit has been introduced, serving for binocular visual observation of the sample or its projection with a magnification of × 7·5 on a screen of 14·6 cm diameter (see § 2.3.3). The Bausch & Lomb base has only a coarse drive and does not allow sample observation in transmitted light. The nose piece carrying Spencer American Optical Co. objectives can only be rotated after removal of the plug-in type electrode holder. There is no facility for electrode projection. A simple coordinate stage allows correct sample positioning. An additional filter with a transmission of 50 per cent can be brought into the light beam to attenuate the laser radiation if

---

FIG. 12. Optical diagram of LMA 1 microscope.

1 Elliptical reflector; 2 Flash tube; 3 Laser resonator rod; 4(*a*) and (*b*) Telescope system of optical adaptor; 5 Diaphragm wheel; 6 Three selector units: deviating prism and field lens for photomicrography, deviating prism and field lens for visual observation, beam splitter cube and field lens for resonator alignment by autocollimation; 7 Analyser; 8 Whole- and quarter-wave plates; 9 Binocular eyepiece tubes; 10 Two objectives in selector slide; 11 Condenser with aperture stop; 12 Wide-angle lens (swing-in); 13 Polarizer; 14 Lenses of illumination system for transmitted light; 15 Optical filter; 16 Field lens for transmitted light with illumination field stop; 17 Three position selector: transmitted light, electrode projection, reflected light; 18 Field lens for reflected light; 19 Projector lamp; 20 Electrode projection beam; 21 Optical filter; 22 Aperture stop; 23 Lenses of illumination system for reflected light; 24 Illumination field stop; 25 Polarizer; 26 Photomicrography unit MF (main body); 27 Projection objective of photomicrography unit MF.

necessary. Furthermore, five aperture stops are available to regulate the intensity according to the requirements of any given analytical problem; they are openings in a diaphragm wheel with diameters stepped from 15 to 125 mm. To change the setting, eighteen screws have first to be removed that hold the metal cover of the laser head.

LOMO, Leningrad, introduced their model MSL 2 at the Belo-Russian Scientific-Technological Conference on Spectrum Analysis, Minsk, in the Autumn of 1967.[72] Like the Zeiss model LM this microscope allows visual observation by the bright field and polarization method either in reflected or in transmitted light. The prototype has three objectives: two flat-field achromats, namely × 5/0·1, focal length 40 mm, working distance 24·9 mm, and × 8/0·2, focal length 25 mm, working distance 18·8 mm; and one achromat × 14·4/0·3, focal length 13·9 mm and working distance 5·7 mm. The eyepiece pairs for the binocular tube unit have magnifications

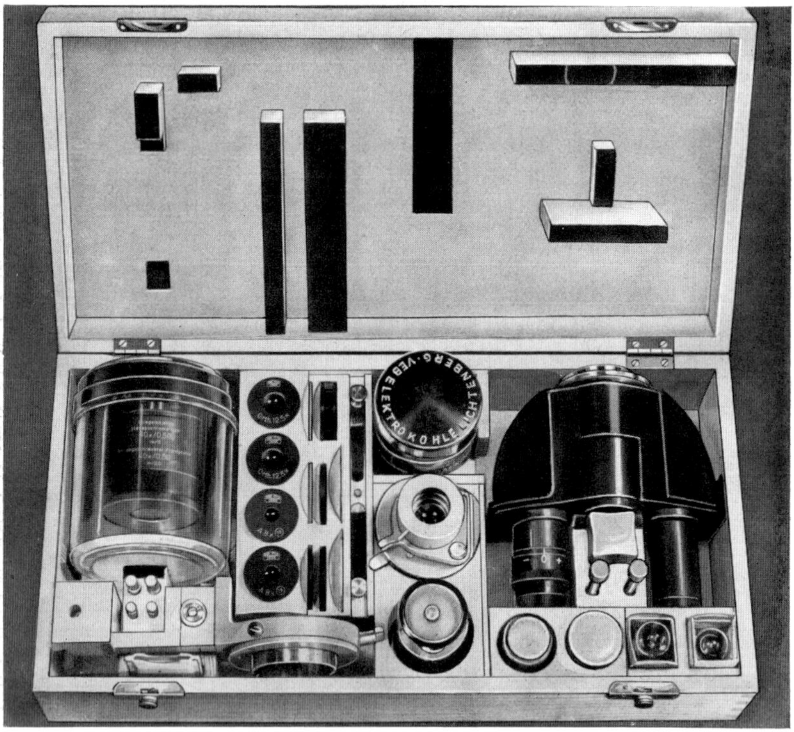

FIG. 13. Case with LM microscope outfit (VEB Carl Zeiss, Jena) showing interchangeable binocular eyepiece body, the three objectives, two pairs of eyepieces, condenser tray, condenser for transmitted light, filter set and various other accessories and spare parts.

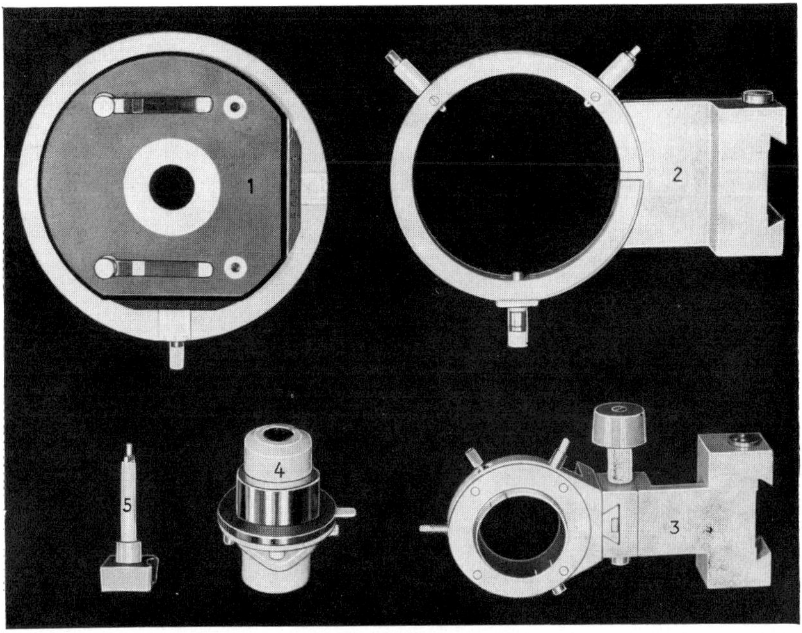


FIG. 14. Microscope stage and condenser for transmitted light and mounts for LM microscope base:

1 Mechanical stage; 2 Stage mount; 3 Condenser tray with focusing knob and clamping and adjustment screws; 4 0·6 Condenser with wide-angle lens; 5 Key.

of ×10, ×15, ×20 (compensating eyepieces) and ×28·5 (orthoscopic). The coordinate stage of the microscope has a 20 mm and a 40 mm range of motion in the $x$ and $y$ directions and can be moved up and down through 50 mm. The electrode holder is adjustable and especially designed for taking thin metal electrodes. The quartz condenser of 77 mm focal length is fitted directly to the microscope in a sliding sleeve.

Two French laser microscopes have to be mentioned for their novel design. Since 1963 M. Loquin,[69a] Paris, has offered the models 'Stabifocal' and 'Néofocal' with a built-in ruby resonator of 7·5 mm diameter. The Stabifocal laser, a research instrument, is bridge-shaped and has a rotating nosepiece with six objectives (Fig. 15a). The laser radiation emerges at right angles to the plane of the sample stage and is directed into the microscope along the optical axis by means of two deviating prisms. The model Néofocal has to a large degree been rendered automatic for industrial applications; it has a revolving wheel nosepiece with three objectives (Fig. 15b). A television camera can be fitted to both Loquin microscopes, which are also classified as photon instruments, without any additional

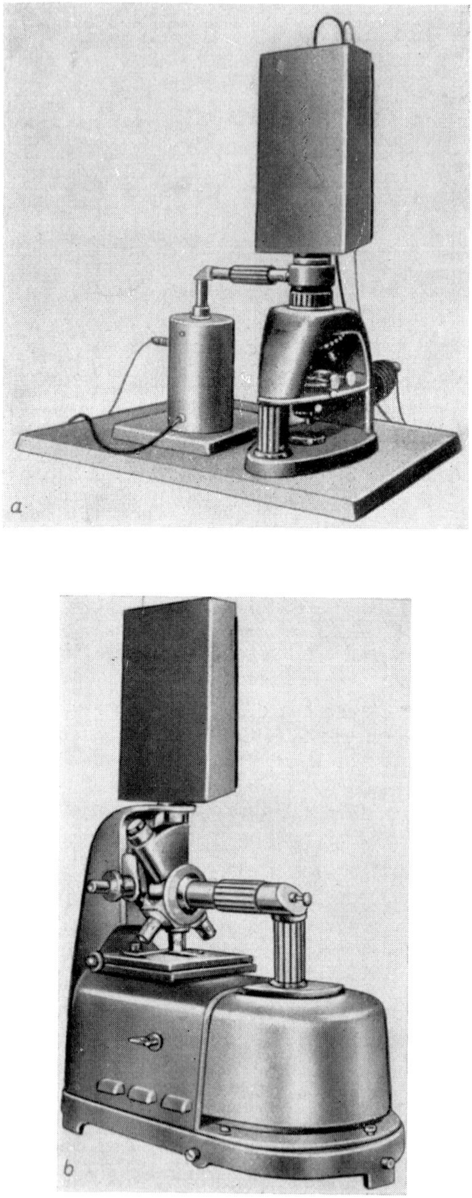

FIG. 15. Laser Microscope of Loquin, Paris.
(*a*) Model Stabifocal. (*b*) Model Néofocal.

support. In the opinion of the manufacturers accidents in experimenting with the laser radiation can be avoided in this way. Such an effort, however, does not seem to be necessary for micro-spectrochemical analysis. A system of electrical safety switches is called for (see the description of the Zeiss LM model).

Bruma and Velghe[18, 19] have converted a commercial version of the Stabifocal model for work with transmitted light into a simple laser microscope by adding an attachment with two adjustable plane mirrors which can be set in two positions, for 'sample photography' and 'laser irradiation'. When the mirrors are swung out of their working positions the sample can be viewed visually by means of a further observation attachment. Besides this adaptation, Bruma and Velghe have taken out a patent for the combination of a profile projector for observation in reflected and

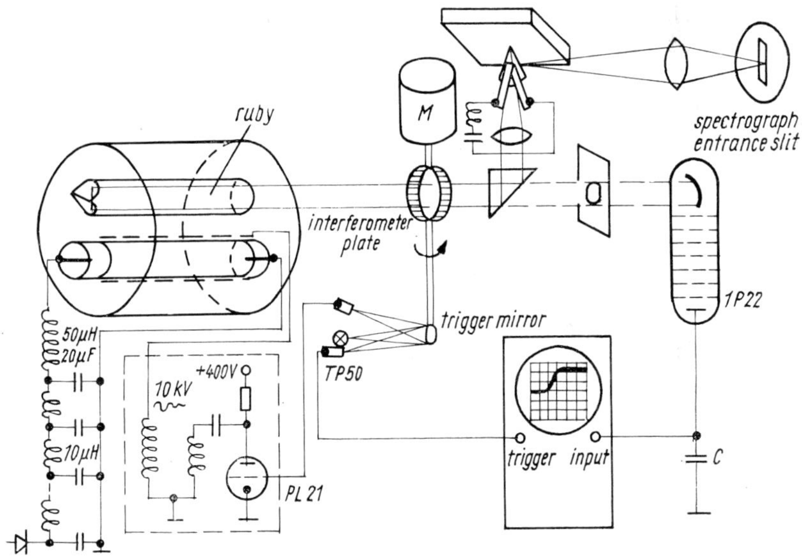


FIG. 16. Laser outfit for spectrum analysis constructed by Institute for Spectrochemistry, Dortmund (G.F.R.).

transmitted light with a solid-state laser in a resonator housing for straight or helical flash tubes.

Hagenah, Laqua and Klocke of the Institute for Spectrochemistry, Dortmund, G.F.R., developed a laser outfit for spectrochemical analysis with an optical system of the Le-Chatelier type. The sample and the auxiliary electrodes are placed above the microscope objective; a coverglass protects the objective against condensation and damage by ejected sample particles. A special feature of this outfit is the possibility of measuring the laser energy on an oscillograph by directing, through a partially reflecting prism, some of the radiation on to a detecting photomultiplier type 1 P 22 (Fig. 16). In an improved version the deviating prism was dispensed with and the elliptical reflector, the flash tube and the ruby rod with external mirrors were arranged vertically. As the visual sample observation and the focusing of the laser radiation are carried out with different objectives, a high precision nosepiece which takes three objectives had to be constructed by the Institute. The new model also has an autocollimating telescope, a high-speed motor for rotating the interferometer plate and a motor-driven scanning device for macro-analysis with meander motion over a range of $2 \times 2$ cm in steps from 20 to 30 μm. A chamber for excitation in special gases can be attached.

The laser microscope of Maser Optics Inc. likewise has an inverted stand. It is intended for experiments with active and passive Q-switching, or with un-switched and also with high repetition rate lasers.

### 2.3.1. Microscope objectives

In the various methods of laser micro-spectrochemical analysis (see § 3.1) the laser radiation, coherent in time and in space, is concentrated on the sample surface by means of an objective. As the real spectroscopic source, namely the auxiliary spark gap, is also situated between the sample and the objective a system with a very large work distance is required. Furthermore the optical elements have to be shielded against sample particles spattered during vaporization. Such microscope objectives have previously been developed for high-temperature microscopy.

The working distance of the Nf flat-field achromatic, non-immersion systems for work with reflected light is not adequate for micro-spectrochemical analysis.

Objectives with long working distances have been designed as purely refracting and as catadioptric systems; in the latter the light beam is refracted by some and reflected at other surfaces of the system.

The first American microscope designed by Brech and Schuch at Jarrell-Ash Co., USA, had a refracting objective of lateral magnification × 20 and numerical operture 0·25, made by Cooke, Troughton and Simms Ltd., now Vickers Instruments Co., York, England. In their second model, Jarrell-Ash incorporated an air-spaced uncemented objective × 10 with 32 mm focal length;[51] this gives on metal samples a minimum focal spot of 100 μm diameter.

Jarrell–Ash guarantee a laser life of one year or 10000 shots whichever is the shorter, but the guarantee does not cover the microscope objectives. This exception is very significant, indicating possible difficulties with dust particles which can, and often do, lead to serious damage of the lenses after a relatively small number of analyses.[53]

Table 4 gives technical data of the focusing optics of several manufacturers and institutes. Most of the systems are achromats with lateral magnification from ×4 to ×10; one of the exceptions is the reflecting objective ×20 of the prototype of Shimadzu, Kyoto. Jeol, Tokyo, has one refracting objective ×10 in a triple rotating nosepiece.

Riesenberg of VEB Carl Zeiss, Jena, computed the reflecting flat-field apochromatic objective of magnification ×40 and numerical aperature 0·5, corrected for image formation at infinity, incorporating a plane-parallel front plate in a removable screwed-in cell, hence not requiring an

TABLE 4. Specification of mains supply (source) unit and microscope objectives of various laser micro-spectrochemical analysers

| | *Supply unit* | | | *Objective* | | |
|---|---|---|---|---|---|---|
| *Manufacturer* | *Capacitance* (μF) | *Inductance* (μH) | *Voltage* (kV) | *Type* | *Lateral magnification and numerical aperture* | *Working distance* (mm) |
| VEB CARL ZEISS, Jena | 1–2·5 | 30–1000 | 0–5 | Flat-field achromat | ×16/0·20<br>×4/0·05 | 18 |
| | | | | Reflecting flat-field apochromat | ×40/0·50 | 15·7 |
| JARELL-ASH | | | | | | |
| Model 45-601 | 5–10 | 0–155 | 1·5–2 | Achromat | ×10<br>×20/0·25 | 13 |
| Model 45-604 | 25 | 40–100 | 1·5–2 | Achromat | ×10 ($f$ = 32 mm) | |
| LOMO, Leningrad Model MSL 2 | 0·5–1·5 | 15–500 | 8 | Flat-field achromat | ×5/0·10<br>×8/0·17 | 24·9<br>18·8 |
| | | | | Achromat | ×14·4/0·30 | 5·7 |
| SHIMADZU | 10 | variable | 2·5 | Achromat<br>reflecting objective | ×10<br>×20 | |
| CISE | | | | | | |
| Model I | 14 | | 2·4 | Achromat | ×7 | |
| Model II | 14 | | 2·3–2·5 | Achromat<br>Achromat | ×4<br>×32 | |
| JEOL | | | | | | |
| Model JLM 200 | 1–10 | 25, 50, 100, 125, 150 | 0–3 | | ×2·5, ×5, ×40 | |

external cover plate when used in its complete form.* The optical diagram of this Zeiss system is shown in Fig. 17: the objective has two reflecting surfaces, namely the concave mirror (3) and the coated convex surface of the second component of the lens-mirror combination (2). The objective is protected by a plate (1). The lens-mirror improves the image field and the image field curvature and gives the long working distance.

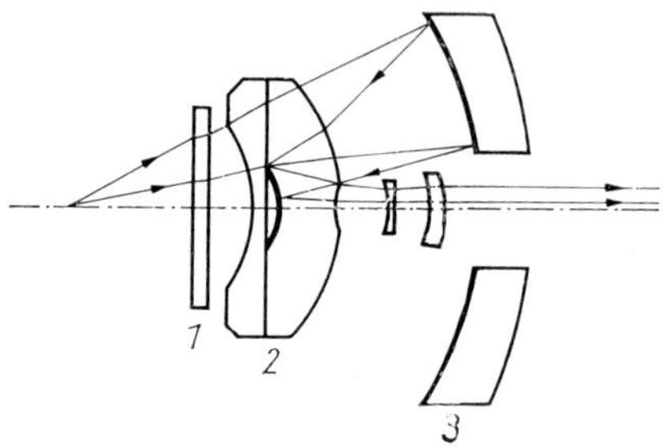


FIG. 17. Ray diagram of microscope lens-mirror objective $\times 40/0{\cdot}5$ of Riesenberg.

Such optical systems have in addition to the external (numerical) aperture limitation also an internal one (central vignetting), in this system 32 per cent linear.

If with the objective of Riesenberg the infra-red laser radiation of wavelength 1·06 μm is to be focused on a sample, the object distance normally set for the wavelength of 550 nm should be reduced by 5 to 6 μm by means of the fine motion control on the microscope stand. The designation as 'flat-field apochromatic' indicates, amongst other things, that the objective absorbs ultra-violet and infra-red radiation as do all objectives of this type which are made of normal optical glasses.

The quartz protecting plate for the reflecting objective $\times 40/0{\cdot}5, \infty/0$ has a thickness of 2 mm $\pm$ 0·04 mm; mounted in its cell it can be screwed on to the objective and is then perpendicular to the optical axis within a tolerance of 12 minutes of arc. With it in position the objective has a working distance of 15·7 mm. The outer diameter of the objective mount is 65 mm; it is tapered towards the lower end so that the rim facing the

* This specification will in the following text be denoted in abbreviated form $\times 40/0{\cdot}5, \infty/0$.

If the objective is used with the cell and its front plate removed, an external cover-plate of 2 mm thickness has to be introduced in the light beam. This mode of operation will be indicated by the symbol ..., $\infty/2$. The same applies to an earlier version of an objective which was not provided with a built-in front plate (see p. 52, Fig. 20).

Finally the symbols ..., $\infty/-$ signifies that the optical system of the objective is not sensitive in this respect and allows working with or without a plane-parallel plate in the lightbeam (see p. 114).

sample has a diameter of 45 mm. When the authors found, in their experimental work, that the carbon electrodes came too near to the objective mount, they as a safety measure replaced the light metal cell of the protecting plate by a plastic one. Without the plate in the light beam the working distance becomes 18·8 mm, which is not suitable for laser microspectrochemical analysis; with carbon electrodes of 5 mm diameter such a working distance is in fact not required (Fig. 18).

FIG. 18. LMA micro-spectrum analyser prototype (December 1964). Electrode holder III between reflecting objective × 40/0·5 and mechanical stage H 5, with sample in position.

If no tube factors have to be taken into account one can work with the × 6·3 and × 12·5 eyepieces in the useful magnification range of × 250 to × 500. The combination of × 0·63 condenser for reflection work, × 40/0·5 reflecting objective, and × 12·5 eyepiece leads to an overall magnification of × 315, which has been found very satisfactory for routine work. The adjustable eyepiece should always have cross lines for setting on the sample and taking readings on the graticule scale. With an appropriately calibrated graticule scale the diameter of the crater formed by the laser irradiation can easily be determined. The authors, using a Zeiss reflecting objective ×40/0·5, successfully carried out several thousand spectrochemical analyses (Fig. 19).

The use of catadioptric systems is only advisable for work with ruby lasers with pulse energies up to about 1 J. Laser energies above this value may damage cemented optical components, coatings and mirror surfaces. Outfits with high power lasers need uncemented objectives or single lenses. The development of such types of microscope objectives is still far from being completed; it will probably lead to the construction of quasi-monochromatic systems for concentrating the photon packets on to the sample. Objectives with two reflecting surfaces of the type described or low power, flat-field achromats in conjunction with eyepieces or a projection screen will then serve for visual observation of the sample.

When testing a commercial American Q-switched laser probe the National Bureau of Standards, Washington, D.C., found that the very high spectral energy density damages the optics of the laser head and the microscope objective.[103]

In focusing the laser radiation of the microprobe in the Q-switched mode on to the sample surface, the optical system ×20/0·25 is very soon

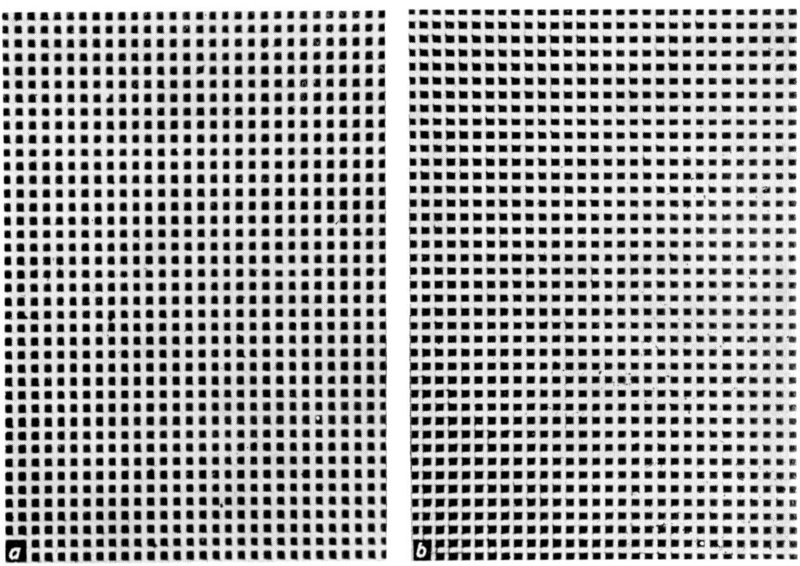


FIG. 19. Photomicrographs of a two-dimensional reflecting grating (spacing 5 μm) taken with a Zeiss reflecting objective ×40/0·5.
(*a*) Test photograph before experiments with focused laser radiation.
(*b*) Test photograph after more than 2000 spectrum analyses with output energies up to 0·5 J (solid-state laser ZFL 750 with neodymium glass resonator).

destroyed.[104] First of all, discoloration of the optical cement was observed; this rendered the system completely unusable after only 100 laser firings, each with several 1 MW spikes. Even with single 4 MW pulse working, only up to 500 analyses could be carried out with one objective.

After these bad experiences experiments were made with an optical system not containing any Canada balsam. The results were fractures in the upper lens components: a higher proportion of the laser radiation with its high energy density of $5 \times 10^7$ W/cm$^2$ was now reflected by the uncemented surfaces and brought to a focus within the upper objective lens, which caused unintentional vaporization of the material. A quarter-wavelength coating reduced the reflection, but insufficiently.

A further drawback of the objective ×20/0·25 of 13 mm working distance is the lack of a protecting plate. A guard against ejected sample particles has to be improvised with polyester films.[104]

The reflecting objective ×40/0·5, ∞/0 has a focal length of only 6·3 mm. This feature is decisive for the size of the obtainable focal spot; the crater diameter was up to 40 μm in the first Zeiss micro-spectrum analyser and up to 10 μm with the model LMA 1 with optical adaptor.

In practical work the spectroscopist often is interested in carrying out, without any special sample preparation, a rapid semi-micro-analysis on a sample area of about 200 μm diameter. Since 1966 a Zeiss objective has been available for such work; it is a flat-field achromat ×16/0·2, of 15·6 mm focal length, with a plane-parallel front plate and with a long working distance of 14 mm. An earlier version of this objective is illustrated in Fig. 20.

Three alternative combinations can be provided with the double objective selector slide of the model LMA 1: large-field objective ×4 and reflecting objective ×40/0·5; or large-field objective ×4 and flat-field achromat ×16/0·2; or reflecting objective ×40/0·5 and flat-field achromat ×16/0·2. In mineralogical routine investigations, very often the authors preferred the last combination of two objectives which both permit focusing of the laser radiation.

In 1965 Hein[50] reported on a Schwarzschild reflecting system ×24/0·65 developed at the Institute for Optics and Spectroscopy at the German Academy of Sciences, Berlin-Adlershof. It has a focal length of 19·5 mm and a working distance of 32 mm. The internal aperture is 0·31. Through central vignetting, 22·5 per cent of the incident radiation is cut off. This microscope is used for investigating the microstructure of electroluminescence at low temperatures (cell windows of 9 mm thickness). The limit of resolution is 2 to 3 μm.

According to the specification this objective could be used for special analyses. In these, energy losses through central vignetting, however, should be avoided, and normally the necessary sample quantity should be vaporized with a single laser shot. The exceptionally long working distance could be an advantage in particular spectroscopic problems arising in work on heavily spattering materials. The mentioned Zeiss objectives ×40/0·5, ∞/0 and ×16/0·2, ∞/0 have been successfully employed in most analytical tasks.

The necessity of using special objectives for laser spectroscopy was shown by experiments carried out at the Institute for Spectrochemistry, Dortmund; in the first investigations with laser radiation at this Institute

FIG. 20. Zeiss flat-field achromat ×16/0·2, ∞/2* with long working distance used with laser micro-spectrum analyser LMA for semi-micro-analysis of a rock sample.

a number of achromats ×10/0·30 of the type incorporated in photoelectric microphotometers were damaged and became useless. The Institute now employs uncemented refracting objectives of magnification ×20 and focal length 8 mm made by Vickers Instruments Ltd, York, England, and other refracting objectives, also uncemented, of 15 mm and 27 mm focal lengths, specially designed by Slevogt in Berlin and made by a firm in Wetzlar, G.F.R.

The monochromatic objective of 15 mm focal length and aperture ratio *f*/1·5 has been described in detail by J. Kross. Two similar systems of this

* See footnote p. 48.

specification were designed in the Optical Institute of the Technical University, Berlin; they are made of the glass types BK 7 and SF 2 and consist of three uncemented lenses and a plane parallel plate in the image space. Small zonal aberrations are an outstanding feature of the SF 2 glass objective.

The relationship between the energy density of focused laser radiation and focal length of the objective is given by the following equation:

$$\text{Energy density} = \frac{E}{\pi r^2} = \frac{E}{\pi \tan^2 \alpha} \frac{1}{f^2}$$

where $r = f \tan \alpha$ = half diameter of focal spot
$E$ = energy
$\alpha$ = the aperture angle.

It follows that for the same divergence of laser resonators the crater diameter – which is not identical with, but often dependent on the focal spot diameter – depends on the focal length of the microscope objective.

A fairly wide aperture also is desirable.

### 2.3.2. Laser optic

What is meant by 'laser optic'? Statements in American catalogues such as 'the complement of lenses usually available with the typical microscope is not included' indicate that evidently microscope objectives are not considered to be part of 'laser optic'. This comprehensive general term only covers

The interchangeable resonators (crystals or doped glass)
The xenon flash tube or tubes
The telescope systems with divergence stop interposed between resonator and entrance pupil of the objective
The safety shutters in the laser head and in front of the eyepiece tube
The beam splitters with dielectric coatings
The positioning lens
The field lens

As far as can be judged from the literature the optical systems are achromatic. No doubt in the future many improvements will be achieved in this branch of applied optics.

The optical adaptor of the model LMA 1 that can be fitted between the LM microscope and the resonator head is shown in Fig. 61. The effect of the adaptor is essentially that of a telescope imaging the laser beam at unit magnification. A reduction of the focal spot down to a diameter of a few micrometres at the cost of total energy is obtained by means of stops in the focal plane of the two telescope lenses which are imaged in the sample plane. Three alternative diaphragms are provided in a rotating mount, numbered 1 to 3 in descending order of size.

### 2.3.3. Accessories of laser microscopes

Usually photomicrography equipment is not included in basic laser microprobe outfits. Zeiss provide two photomicrography outfits, one with a camera attachment of 24 × 36 mm, the other with a camera attachment of 6 × 6 cm.

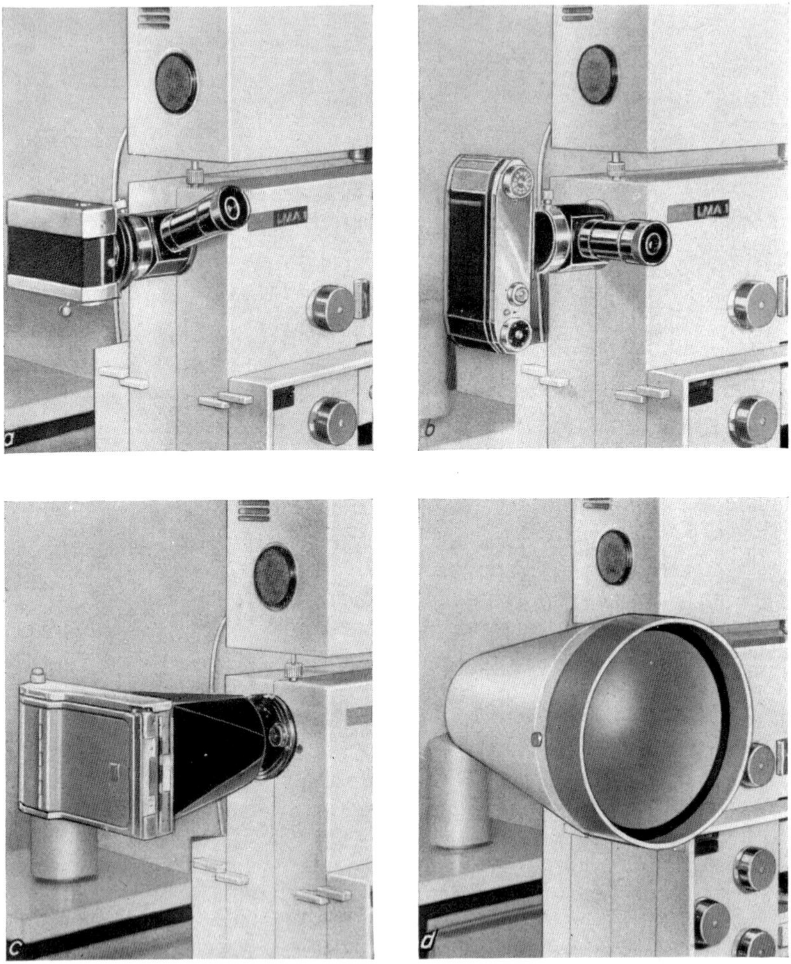

FIG. 21. Microscope LM.
(*a*) and (*b*) Photomicrography unit MF; (*c*) Polaroid camera; (*d*) Visual observation unit.

In 1966 Jarrell-Ash added a viewing attachment to their second microprobe. This attachment fits on the trinocular tube where alternatively a polaroid camera can be mounted. Apart from educational applications of laser outfits, the practical value of a viewing attachment is debatable. In the majority of cases unprepared samples are analysed with a laser probe and hence the possibilities offered by a viewing attachment are not fully exploited. As shown in Fig. 21 the Zeiss model LMA 1 also takes a viewing attachment and a Polaroid camera, but these two units are not included in the standard outfit. The miniature camera model 'Werra' of the VEB Carl Zeiss can be used with the LM stand (Fig. 21).

The sample stage micrometers with 1 mm range and 0·01 mm divisions for use with transmitted and reflected light are also not part of the standard outfit of model LMA 1.

## 2.4. LIGHT SOURCES FOR SPECTROSCOPY AND THEIR SUPPLY UNITS

### 2.4.1. The two-step method (cross excitation)

Laser micro-spectrochemical analysis is a valuable addition to the established methods of emission spectroscopy of trace elements and major constituents. It no doubt greatly helps in the introduction and routine use of the appropriate new equipment if standard spectrographic outfits without modification of the basic units can be used, together with the well-tried condenser systems aligned on the optical bench in front of the entrance slit. Prism and grating instruments with effective aperture ratios from *f*/11 to about *f*/35 are regarded as standard outfits and are usually found in analytical laboratories.

When adopting laser excitation one has to take into account the effective aperture ratio, the speed, the resolving power, and the dispersion of the commercially available instruments. Basically it is possible, especially with Q-switching, to use as spectroscopic light source the microplasma that is generated when the focused laser radiation strikes the sample. This has in fact been done partly with good results, even if only after modification of the condensing optics and sometimes with considerable difficulties. (Examples will be discussed in § 3.1.)

In fundamental spectroscopic research, and especially in plasma physics, it is essential to direct the radiation from the microplasma itself immediately into the optical system of the spectrograph. The first test results were reported in 1964.[7]

Commercial laser probes with an electrical spark gap between the sample and the front lens of the microscope objective have recently become established for purposes of production control and routine analysis with solid-state lasers as energy source. The focused laser radiation here is primarily used for vaporizing the sample material. The produced microplasma, expanding, according to the laser working mode, at a rate from $10^4$ to $10^6$ cm $s^{-1}$,[85,96,118] contains the chemical elements of a sample particle to be analysed. The elements are excited to emission by a single capacitor discharge across the spark gap. Compared with the method of simultaneous sample vaporization and excitation (the microplasma produced by a Q-switched pulsed laser) the analytical sensitivity by this two-step method with 'cross excitation' is increased by a factor of fifteen to twenty.

The sequence of events in the two-step process is illustrated in Fig. 22. The xenon flash tube emits incoherent pumping radiation (chain-line curve) causing laser activity which can, in the discussed case of an unswitched resonator, be demonstrated on an oscillograph as a series of spikes given off in a time interval of a few hundred micro-seconds.

The first particles in the ascending vapour cloud arrive in the space between the charged electrodes in 5 μs. The peak intensity of the microplasma (dotted curve) is reached in 100 to 150 μs. The spark discharge

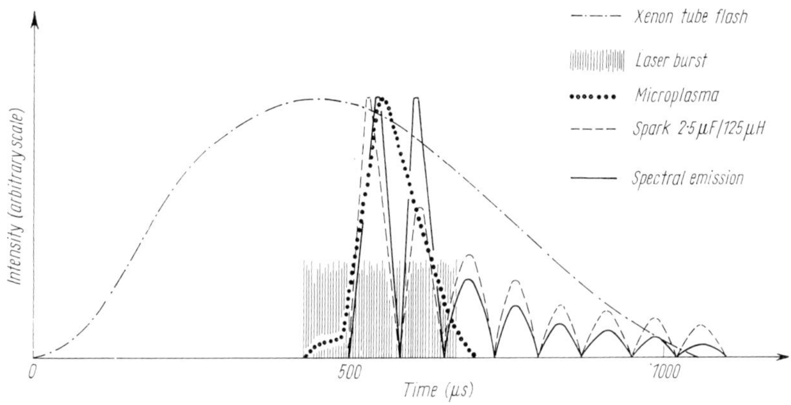


FIG. 22. Time sequence of events in a two-step working procedure with an LMA outfit. Pumping flash, laser burst, microplasma without cross excitation, spark discharge and spectral emission of microplasma excited by this discharge.

between the microscope objective and the sample surface is released between 5 and 100 μs after the start of the laser emission.

The electrical parameters of the auxiliary excitation illustrated in Fig. 22 were 2·5 μF and 125 μH (model LMA 1[87]). The duration of the spark discharge can be altered through appropriate choice of the parameters. Spectrum lines can be determined immediately after the start of the discharge.

### 2.4.2. Set-up of the earlier experimental outfits

A diagram of the first experimental set-up of Zeiss is given in Fig. 23. The points of two carbon electrodes, about 0·2 to 1·2 mm part, are aligned about 1 to 2 mm above the sample surface. A direct voltage of approximately 1·2 to 2·5 kV, which is not sufficient to break down the gap, is laid on across the electrodes. The vapour plume rising from the sample surface after impact of the laser pulse is ionized to such a degree that it causes the break down of the spark gap and a discharge of the capacitor load across it below the front lens of the microscope objective. The plasma of this radiation source consists mainly of vaporized sample substance; its electrode material content is so small that only when electrodes of low purity carbon are used do lines of impurities in the carbon interfere with the sample spectrum; this has often been asserted without experimental proof.

The necessary direct voltage can be obtained from a simple mains supply unit (mains supply unit II – see Fig. 24). No voltage stabilization is required and no setting acuracy for the selected nominal value better than ±25 V. The charge for the spark generation is accumulated in the capacitor $C_3$. The resistance $R_2$ in the charging circuit is chosen so that after

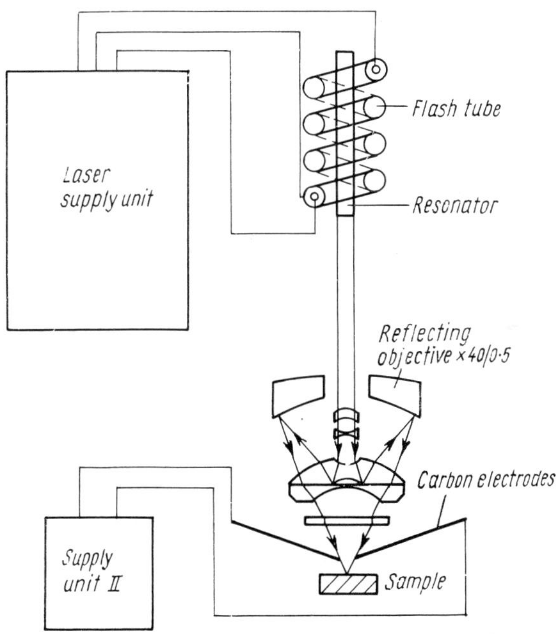


FIG. 23. Diagram of operation of laser micro-spectrum analyser LMA.

the capacitor discharge no arcing can occur between the electrode tips. The charging rate of the capacitor $C_3$ is matched with the working speed of the complete outfit; Krause[13] experimentally established a value of 10 s. The required energy is comparatively low; from the maximum voltage of 5 kV and the maximum capacitance of 3 μF follows its value of 37·5 J. In practical work (see Chapter 3) the appropriate voltage is established in a preliminary test; it depends, as can be expected, on the

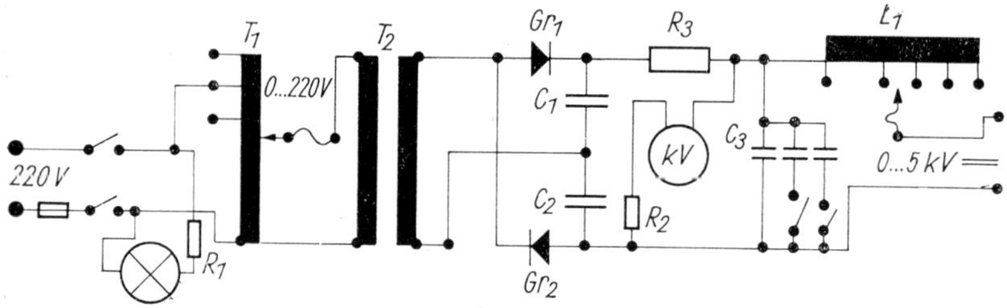


FIG. 24. Circuit diagram of supply unit II of laser micro-spectrum analyser LMA.

electrode separation (Table 5) and electrode shape, and should preferably be somewhat lower than the breakdown voltage. It can be adjusted by means of the transformer $T_1$. It is also good practice to step up the voltage by a factor of two because then the high voltage transformer $T_2$ need only take 2·5 kV, which simplifies the insulating. Short circuiting the instrument inadvertently for some time, for instance through touching both electrodes with a conducting sample (steel, ore, or a similar substance), will not damage the supply unit. The charging resistance $R_2$ is chosen so that it is not overloaded in case there is a short circuit. The transformers and rectifiers too are designed for the appropriate loads. The supply unit model II has selenium rectifiers; it therefore requires no warming up and is ready for use immediately after switching on. The capacitor $C_3$ can be adjusted for different working conditions by bringing into the circuit further capacitors in addition to the basic fixed unit of 1 μF.

TABLE 5. Electrode separation and breakdown voltage for the electrodes of laser model LMA 1 (approximate values)

| *Electrode separation* (mm) | *Breakdown voltage* (kV) |
|---|---|
| 0·2 | 1·5 |
| 0·4 | 2·05 |
| 0·6 | 2·45 |
| 0·8 | 2·8 |
| 1·0 | 3·0 |

Similarly different tappings are provided for the inductance $L_1$ of the supply unit to allow a setting for the best possible type of discharge ('arc-like spectrum'). The range of 30 to 1000 μH in five to six steps proved to be appropriate. Because of their form, arc-like discharges between pointed carbon electrodes at inductances of more than 500 μH really require a spectrograph with a horizontal entrance slit (Figs. 25 and 26).

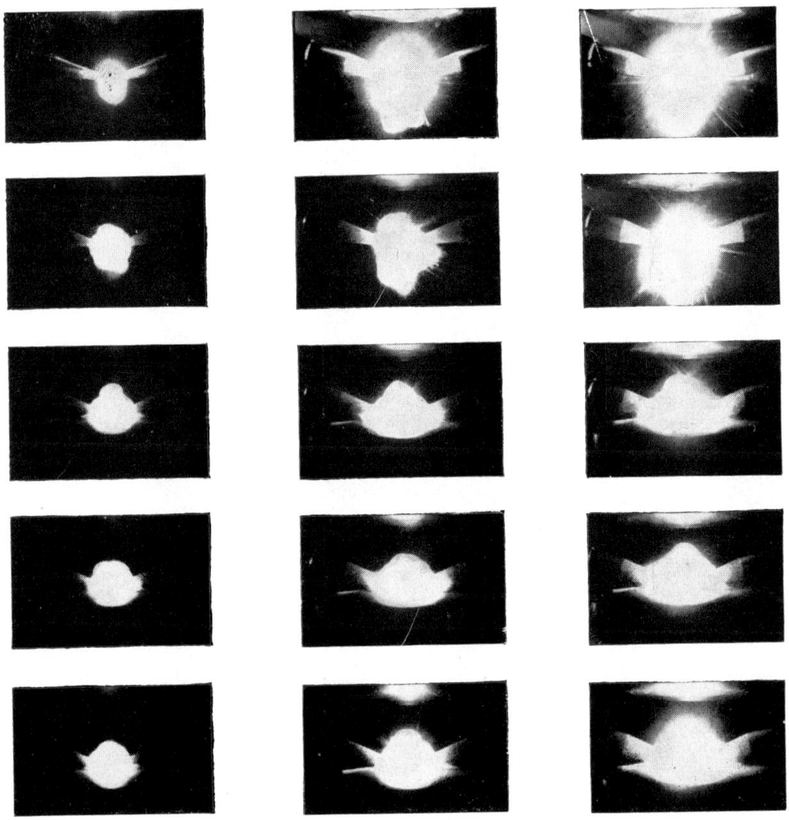

1
2
3
4
5

1 μF 2 μF 3 μF capacitance

FIG. 25. Electrode discharge in laser micro-spectrochemical analysis.

Inductance: 0. Capacitance: variable.
1 Iron electrodes without sample and without laser radiation; 2 Carbon electrodes without sample and without laser radiation; 3 Carbon electrodes with sample and without laser radiation; 4 Carbon electrodes with sample and with laser radiation ($Nd^{3+}$ glass resonator, wavelength 1060 nm); 5 Carbon electrodes with sample and with laser radiation (ruby resonator, wavelength 694·3 nm).

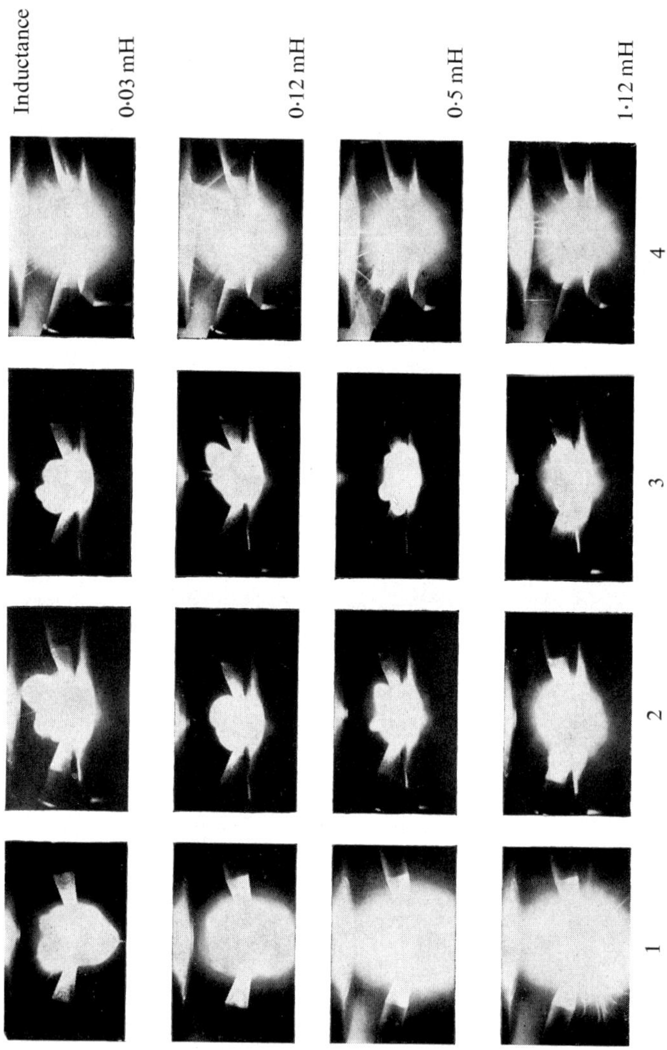


FIG. 26. Electrode discharge in laser micro-spectrochemical analysis.

Capacitance: 2 $\mu$F. Inductance: variable.

1 Without sample and without laser radiation; 2 With sample and without laser radiation; 3 Chromite sample with laser radiation (wavelength 1060 nm); 4 Chromite sample with laser radiation (wavelength 694·3 nm).

### 2.4.3. Specification of mains supply units for commercial and research laser micro-spectrum analysers

*Model LMA 1 of VEB Carl Zeiss, Jena*

In routine production models the mains supply units for the laser and the auxiliary electrodes with their controls and safety devices are housed in one console. The supply and operating components for the auxiliary electrodes of the microscope LM are placed in the third chassis of the movable supply unit of the total LMA 1 outfit (Fig. 5, part 2); included are the charging unit for the five capacitors for the auxiliary spark gap, the inductance for the discharge circuit and the supply for the microscope illumination. A complete block diagram is shown in Fig. 27. The main switch of this chassis and that of the microscope illumination, with their indicator lights and fuses, are fitted at the bottom right-hand side of the front panel. Immediately above is the electrode voltage control, continuously adjustable between 0 and 5 kV. The voltage is indicated on the meter on the left side of the chassis panel. On the top left is a selector switch for the discharge capacitance with steps of 1, 1·5, 2 and 2·5 μF. With the second large switch, on the bottom left, the discharge inductance is set, independently of the available four capacitance steps, to one of the following six values: 30, 60, 125, 250, 500 and 1000 μH.

In collaboration with W. Quillfeldt, the authors have carried out investigations to find out if an extension of these setting ranges of the basic standard model LMA 1 is desirable.[85] In this work the LMA 1 microanalyser was used with the plane grating spectrograph PGS 2, fitted with a photomultiplier type M 12 FC 52 instead of the photographic plate, and connected to a pulse oscillograph with a broad-band amplifier. Cadmium sulphide was chosen for the test sample because cadmium, on account of the low number of spectrum lines, is very suitable for these investigations.

The following measurements were carried out:

1 Change of current and voltage during the discharge as functions of inductance and capacitance values
2 Rate of spreading of the microplasma
3 Lifetime of the cross-excited microplasma
4 Intensity change during the spark discharge; analysis with high time-resolution, with emphasis on the initial stage
5 Time curve of generation of the different spectrum lines
6 Effect of change of distance between sample and spark discharge on spectrum line intensity
7 Time curve of initiation of spark discharge across the microplasma

The following results were obtained:

1 The distance between the sample and spark gap has no effect on the moment of start of the spark discharge. The distance between sample

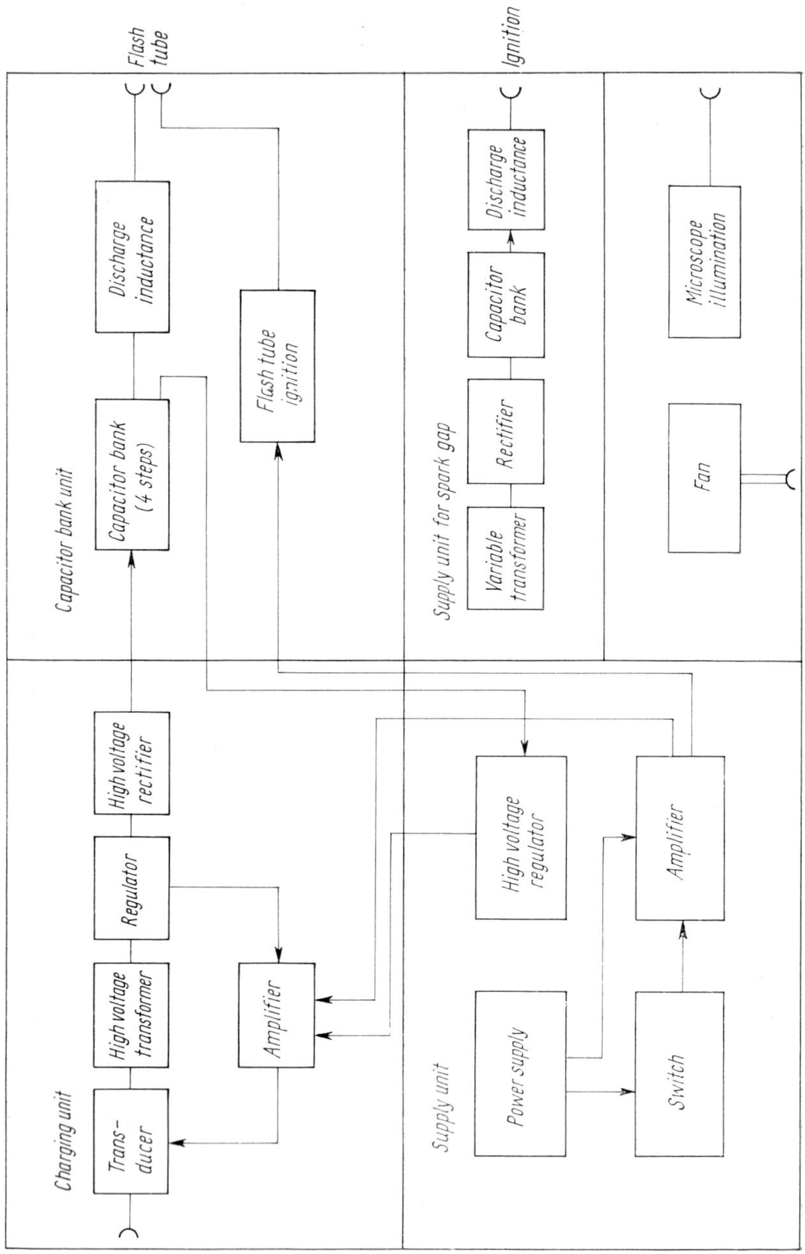


FIG. 27, Block diagram of the LMA 1 supply unit.

and the electrodes was changed from 0·20 to 1·00 mm in steps of 0·20 mm. No change of the moment of start of the discharge could be detected nor any effect on the total intensity of the spectrum lines.

2 A change of voltage across the spark gap, with the electrode separation kept constant, also has no effect on the moment of start of the discharge. Originally such an effect had been assumed. The voltage across the spark gap was therefore varied in steps of 0·5 kV between the limits of 2 and 3·5 kV, the breakdown voltage being 4 kV. But no measurable change of the moment of start of the discharge could be detected. The characteristics of the spark gap only act, through current and discharge duration effects, on the spectrum line intensity.

3 It is not advisable to use simultaneously minimum or maximum inductance and capacitance values, as for instance 30 μH and 1 μF, or 1000 μH and 12·5 μF. When a high capacitance, say of 6·5 μF or more, is used with a low inductance of 60 μH or less the special microscope optics may be damaged, especially at magnifications of × 320 and × 500. In work with 1 μF and 30 μH the time of the first half-period coincides with the low intensity initial phase of the microplasma of 50 μs; hence the intensity of the spectrum lines is not very great. This is also the case for the combination of values of 1 μF with 125 μH and of 2·5 μF with 30 μH.

With high inductance and capacitance the intensity of the spark discharge during its first half period is lower than in the following half periods. With 22·5 μF and 1000 μH the amplitude of the first half period is only 20 per cent of that of the second half period. Therefore no high spectrum line intensity can be expected under these working conditions.

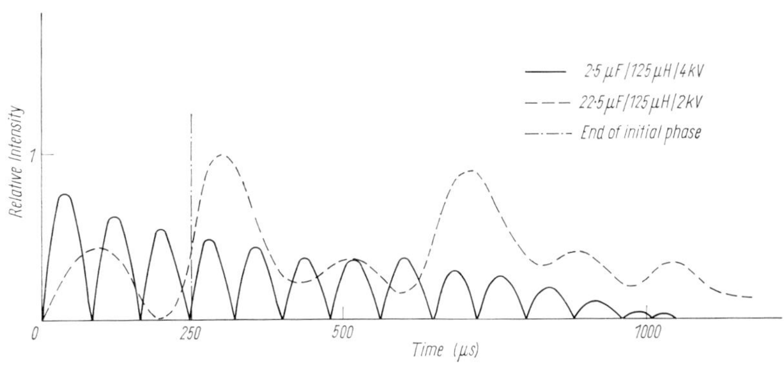


FIG. 28. Intensity curve of spark discharge with 2·5 μF, 125 μH, 4 kV (settings of standard outfit LMA 1) and with 22·5 μF, 125 μH, 2 kV.

4 With synchronization of plasma generation and cross excitation by the auxiliary electrodes, spectra of the same intensities can be obtained with different source unit parameters, as for instance with 2·5 μF, 125 μH, and 4 kV instead of 22·5 μF, 125 μH and 2 kV.

Fig. 28 shows the time curve of the spark discharge intensity for 2·5 μF, 125 μH, and 4 kV, the values of the model LMA 1, and for 22,5 μF, 125 μH, and 2 kV. The ratio of the calculated total intensity values is 1 : 2.

It is interesting to note that during the initial phase of 250 μs in

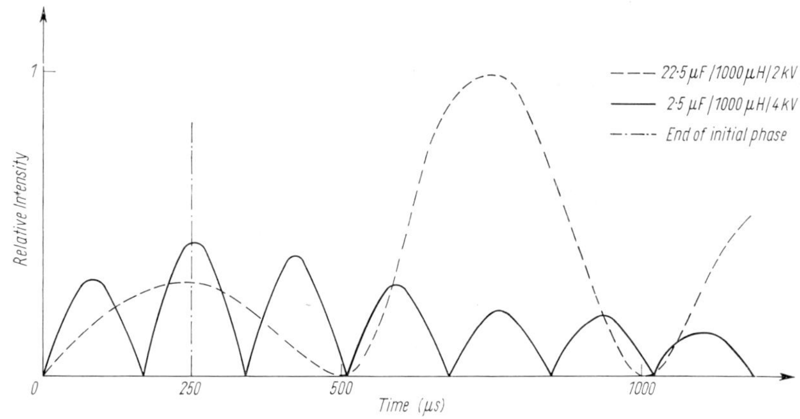


FIG. 29. Intensity curve of spark discharge with 2·5 μF, 1000 μH, 4 kV (settings of LMA 1 outfit) and with 22·5 μF, 1000 μH, 2 kV.

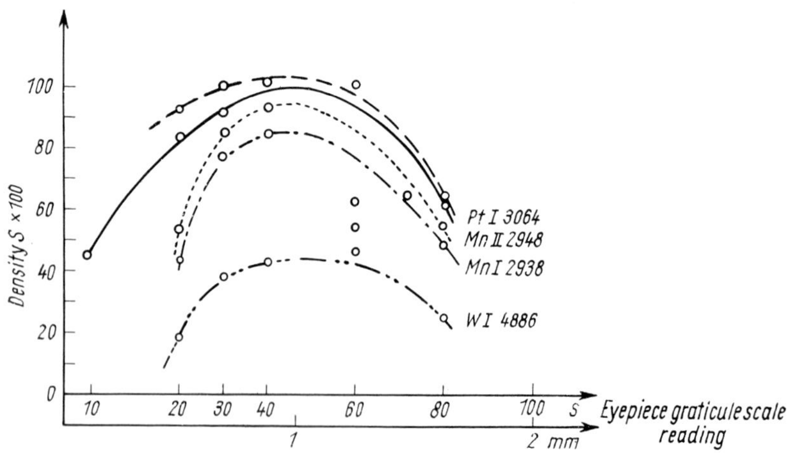


FIG. 30. Change of spectrum line density with change of electrode gap separation.[101]

which about 60 per cent of the total line intensity are generated, the spark discharge intensities ratio is reversed, namely 5:3. This explains why the spectrum intensity does not increase with increased capacitance and lowered voltage. The same relationships of the spectral intensities were observed for the two sets of parameters 2·5 μF, 1000 μF, and 4 kV, and 25·5 μF, 1000 μH, and 2 kV (Fig. 29.). One reason of course is the quadratic relationship between capacitor energy and voltage.

In further investigations of the excitation conditions with the model LMA 1, Quillfeldt[101] triggered off the auxiliary spark excitation with external ignition. This allows the electrode separation to be set independently of the driving voltage or, in other words, to work with high voltages and small electrode separation. The change in the density of different spectrum lines with electrode separation is illustrated in Fig. 30. In general the best electrode separation is about 1 mm.

In the subsequent experiments a delay stage was connected with the external ignition, thus allowing the moment of ignition to be varied. Figs. 31 and 32 illustrate the effect of the change of the interval between laser pulse and spark discharge for platinum and boracite in the period from 200 μs before to 1 ms after start of the laser pulse. Boracite with its lower melting point has its intensity maximum at 0 μs, platinum with its higher melting point at 60 μs.

For a study of the effect of the excitation energy of the auxiliary spark gap on line intensity the electrode separation, the capacitance of 2·5 μF and the inductances of 125 μH were kept constant and the voltage was varied between 2 and 6·3 kV, and then, at constant capacitance and inductances of 6·5 μF and 60 μH, between 1·2 and 4 kV. The initial and final values of the voltage ranges corresponded to approximately equal field energies of the capacitors. The voltage dependence of the spectrum line densities is illustrated for platinum in Figs. 33 and 34. In the first of these illustrations the platinum line 306·4 nm has for comparison been recorded with and without external ignition.

Only in a few cases could differences be found between spectra taken with high capacitance and low voltage on the one hand and with low capacity and high voltage on the other, showing high voltage to be preferable (Figs. 35 and 36).

Spectra of wolframite ($FeMnWO_4$) are shown in Fig. 37. Nos. 1 to 4 were produced with 2·5 μF and 4·5 to 6 kV, nos. 7 to 10 with 6·5 μF and 2·5 to 4 kV. Their intensities are comparable, as can be seen from the illustrations. Most of the investigated spectra show for a capacitance of 25 μF and an inductance of 125 μH maximum density with 5·5 to 6 kV. This gives about 40 J, approximately the value of the original LMA model (37·5 J)[13] which again proves the instrument's design to be the best possible, with the shape only having needed modification.

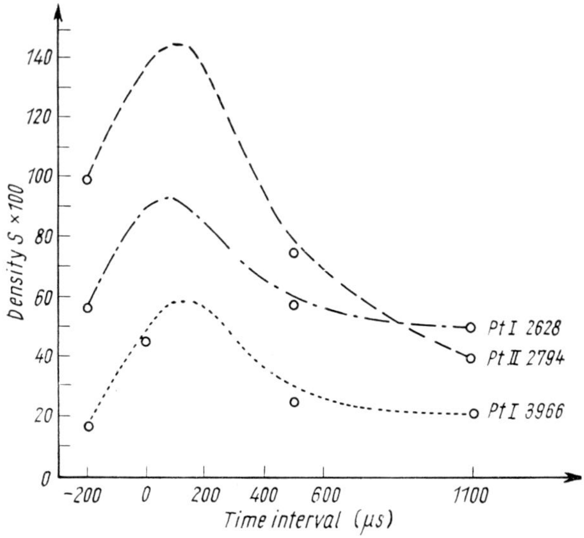


FIG. 31. Change of density of platinum spectrum lines with change of time interval between laser pulse and auxiliary electrode discharge.[101]

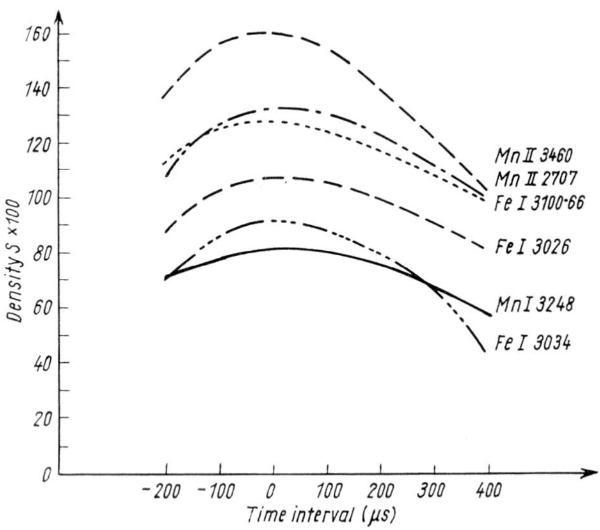


FIG. 32. Change of density of manganese and iron lines in boracite spectrum with change of time interval between laser pulse and auxiliary electrode discharge[101]

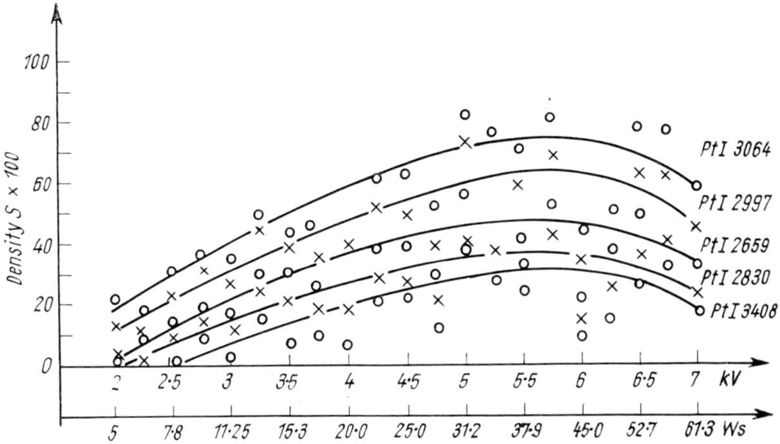


FIG. 33. Change of density of platinum spectrum lines with change of electrode voltage (crater diameter 50 μm, capacitance 2·5 μF).[101]

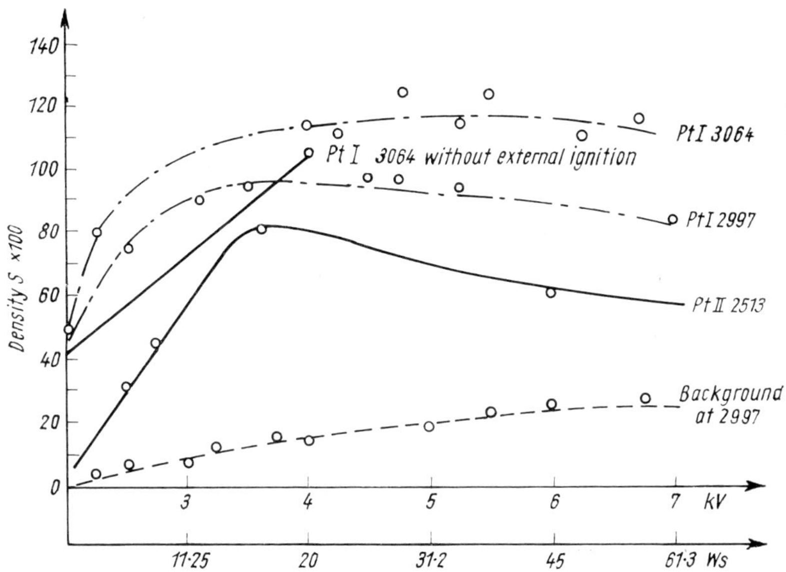


FIG. 34. Change of density of platinum spectrum lines with change of electrode voltage (crater diameter 120 μm, capacitance 2·5 μF)[101]

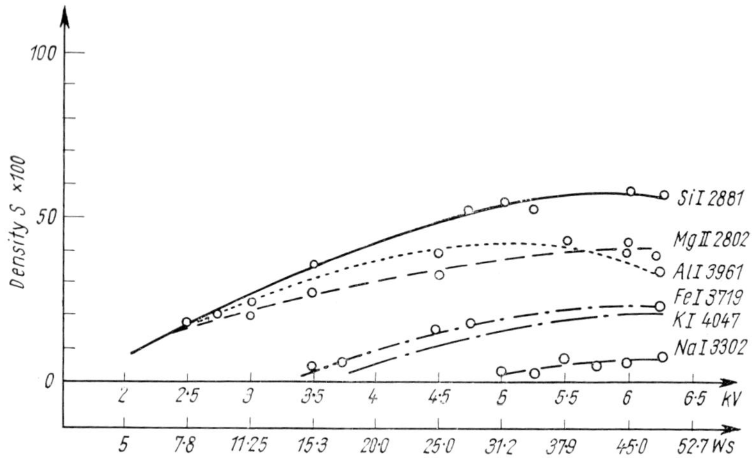


FIG. 35. Change of density of spectrum lines of Si, Mg, Al, Fe, K and Na with change of electrode voltage.[101]

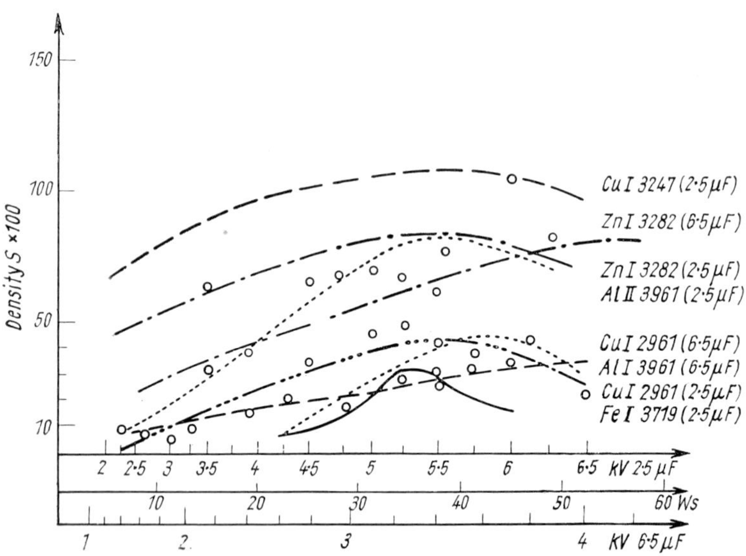


FIG. 36. Change of density of spectrum lines of Cu, Zn, Al and Fe with change of electrode voltage at 2·5 and 6·5 μF.[101]

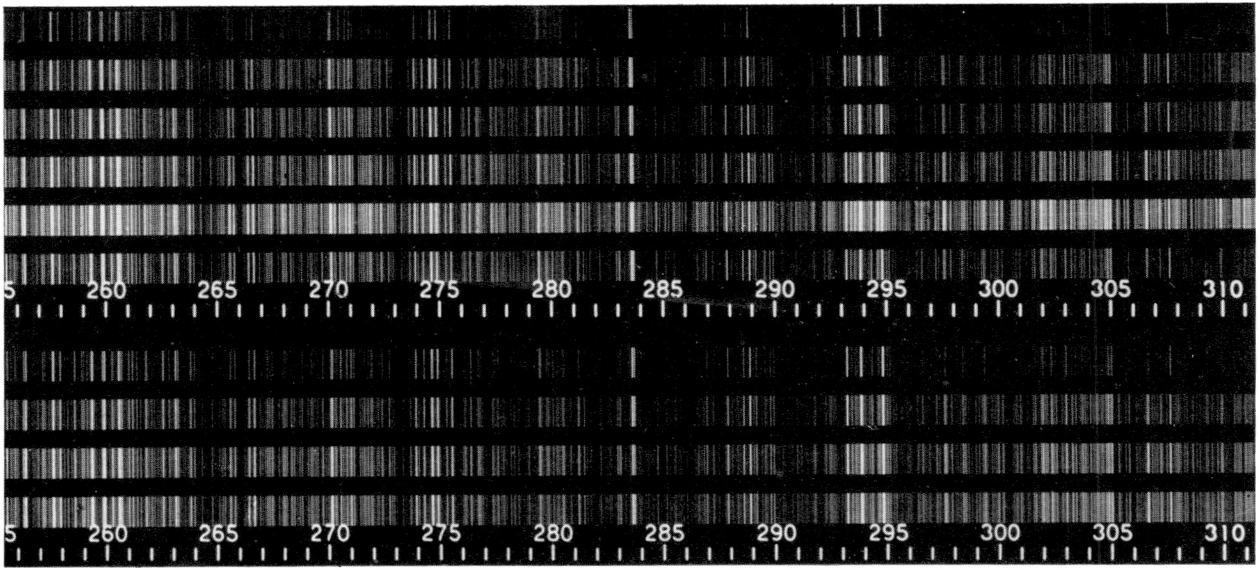


FIG. 37. Spectra of wolframite.[101]

Spectrum 1 2·5 μF, 4·5 kV
2 2·5 μF, 5·0 kV
3 2·5 μF, 5·5 kV
4 2·5 μF, 6·0 kV

Spectrum 5 2·5 μF, 2 × 6·0 kV (two spark discharges after one laser firing)
6 2·5 μF, 4 × 3·0 kV (four laser firings, each causing one spark discharge)

Spectrum 7 6·5 μF, 2·5 kV
8 6·5 μF, 3·0 kV
9 6·5 μF, 3·5 kV
10 6·5 μF, 4·0 kV

*The Jarrell-Ash laser microprobe*

Jarrell-Ash has brought out two models, the first type 45-601 and a new version Mark II model 45-604. In both the electrode voltage can only be set in steps, in the new model of 1·5, 1·75 and 2·0 kV. The electrode holder, of the push-fit type, has two Teflon covered sleeve mounts for the electrodes, National Carbon L-4036. There is no fine adjustment for the electrodes, nor electrode projection. The spark gap is set to 2 mm by means of a spacing jig.

The maximum capacitance of the first model was 10 μF.[104] In the new model Mark II which has an inductance of 100 μH, the capacitance was increased to 25 μF,[53] the basic outfit being fitted with low power microscope objectives that give crater diameters of at least 100 μm or more.[53] At such high capacitance values the CN bands are of course also more strongly excited.

For Q-switch working with the first model 45-601 the following operating ranges were found suitable for spot analyses of low alloy steels: 1·5 to 2·0 kV; 5 to 10 μF; 0 to 155 μH and 0 to 10 Ω.[105] The National Bureau of Standards, Washington put forward a still further specialized recommendation for the spot analysis of metal samples with single pulse laser working and 2 kV maximum auxiliary electrode voltage, namely: inductance of 40 μH and capacitance of 10 μF.

*Model MSL 2 of LOMO, Leningrad*

This 1967 prototype from the USSR also has the supply unit and circuit for the auxiliary electrodes incorporated in the main electronic console for the complete outfit. A system of external ignition for the cross excitation had been developed to allow working with a voltage of 8 kV even with a spark gap as small as 0·5 to 0·8 mm. Three alternative capacitance steps of 0·5, 1·0 and 1·5 μF and three inductance values of 15, 100 and 500 μH are available. The discharge is controlled through the laser pulse by means of a trigatron. The maximum duration of the auxiliary spark discharge is about 500 μs. This Soviet instrument was designed for excitation of spectrum lines of the excitation level III, as for instance Cl 319·14 nm and P 323·36 nm.

With the lowest and highest inductance values the discharge current has the form of a double pulse. For the excitation of metalloid spectrum lines in the visible range the first of the pulses, of about 50 μs duration, giving momentarily more than 2000 A, is decisive. The second current pulse gives 200 A and has a duration of 400 to 500 μs.

The very different specifications of the various standard production and research instruments can be judged with the help of Table 4 (p. 47). From the included technical details of the microscope objectives used, an interrelationship becomes apparent between the chosen maximum capacitance of the discharge circuit and the performance of the focusing system, such as magnification, numerial aperture and working distance. Even with a capacitance of 3 μF the discharge zone is coming dangerously near to the protecting plate of the high performance reflecting objective. For this reason the maximum capacitance of the production model LMA 1 has been fixed at 2·5 μF. The effect of the electrical parameters of the cross excitation circuit can basically be summed up as follows:

Increase of the spectrum intensity – and with it the CN bands – with increased capacitance

Increase of the ratio of spark to atomic spectrum intensity with increased excitation voltage

Decrease of the ratio of spark to atomic spectrum intensity with increased inductance.

It is remarkable that, by varying the electrical parameters of the spark gap, the intensity of the C I, C II and C III lines can be controlled much more than the intensity of spectrum lines of the chemical elements contained in the sample.

## 2.5. CONDENSER OPTIC

There are two possible forms of laser micro-spectrochemical analysis, namely the immediate analysis of the plasma produced at the sample surface and its analysis using an auxiliary spark gap between the sample and objective (cross excitation – see details in § 3.1). In the first method the source is directly imaged on the spectrograph slit, usually by a reflecting condenser system. Such a system with a spherical mirror for imaging the microplasma on the entrance slit, suggested by Debras–Guédon and Liodec[27] for spectrographs with an horizontal slit, is illustrated in Fig. 38.

A system of a condenser lens combined with a plane deviating mirror gave satisfactory results in the laser macro-analysis of steel samples at the Ford Works in the USA. Here too an image of the microsplasma is directly formed in the plane of the slit. The laser radiation is concentrated on the sample surface which makes an angle with the optical axis of the ruby resonator slightly different from 90°. The plane mirror between the condenser lens and the entrance slit is inclined at 45° to the light beam and orientates the beam's axis, which is identical with that of the solid-state laser, parallel to, but slightly displaced laterally from, the optical axis of the Littrow quartz spectrograph ('This allows the spectrograph to 'see' the excitation') (Fig. 39).

In the method with an auxiliary spark gap between the sample and the objective – as in conventional spectrum analysis – either slit illumination for qualitative analysis or intermediate image formation, usually possible only with condenser lenses, can be chosen.

The imaging optic for qualitative analysis also includes an illuminator for electrode projection, a commercial Zeiss product; it has to be mounted on special carriages for the optical bench in place of the spark stand. In the new Zeiss model LMA 1 the electrode projection unit is incorporated in

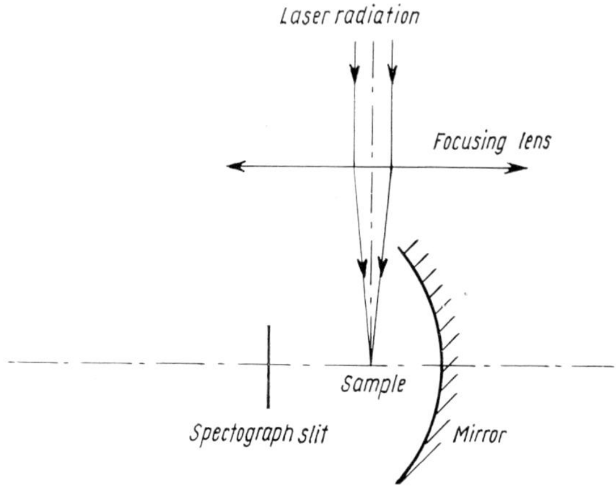


FIG. 38. Imaging the microplasma produced by laser radiation on a horizontal spectrograph slit according to Debras–Guédon and Liodec.[27]

the laser microscope. In addition the authors also used a quartz condenser lens ($f$ = 80 mm) for imaging the fast spark discharge on the spectrograph slit. The quartz protecting plate that has been introduced to prevent damage through metal particles ejected from the sample is not required in laser micro-analysis. The authors found that through placing a suitable spherical mirror behind the microscope in place of the electrode projection unit part of the light emitted in this direction could be utilized for element detection. This, however, does not in general lower the limits of detection, but to a certain extent helps when important findings are to be recorded as enlargements of spectrum plate sections.

The optical lay-out in quantitative laser micro-spectroscopy calls for an intermediate imaging system. Preuss and Strock suggested, for the cathode layer method, systems of pairs of quartz condenser lenses with a diaphragm. In the Zeiss system three single quartz lenses are used: the light source is imaged by the condenser lens ($f$ = 80 mm) with a magnification of about × 3 on a diaphragm disk immediately in front of a second lens ($f$ = 160 mm). Uniform illumination of the slit is obtained because the aperture of the first field lens is imaged by the second single quartz lens in the slit plane. In front of and close to the spectrograph slit the third lens ($f$ = 200 mm) is placed which projects the enlarged intermediate image into the plane of the aperture stop of the collimator objective.

The optimum conditions for filling the aperture of the spectrograph cannot, of course, be met simultaneously in the ultra-violet and the visible spectrum range with simple condensing systems for intermediate imaging. The necessary setting adjustments and the useful rules of thumb are well known in experimental work. In the typical cases of micro-analysis the chemical composition of a sample part under test it not known; in order to avoid duplication of the work a condenser system is therefore desirable that has equal focal lengths in the visible and ultra-violet spectrum regions. VEB Carl Zeiss, Jena, lists two ultra-violet achromats of 75·8 mm and 150 mm focal length, each consisting of a quartz and lithium fluoride lens combina and achromatized for the wavelengths of 253 nm and 589 nm respectively. For the useful wavelength range of the Zeiss spectrograph model Q 24, a single condenser system for intermediate imaging, which fully images light sources up to 5 mm in diameter in the aperture stop of this medium quartz spectrograph is now sufficient.

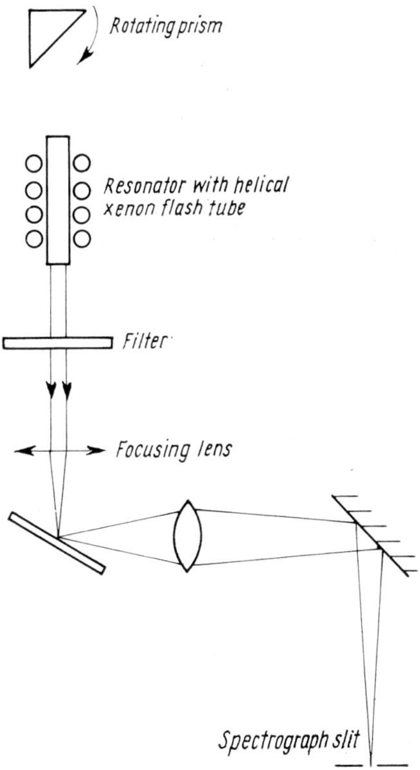


FIG. 39. Imaging optics according to Runge, Minck and Bryan.[113]

Because of the greater distance between light source and spectrograph slit and because of the increased absorption of radiation compared with the condenser optic for qualitative analysis, multiple laser excitation of about five pulses is required in quantitative analysis. For many geochemical problems a semi-quantitative analysis is adequate; neither the described ultra-violet achromats nor any other system for intermediate imaging are employed in such work, but a magnified image of the spark discharge is formed with a single quartz condenser. To avoid the spectrum lines on the photographic plate becoming unilaterally tapered the points of the inclined carbon electrodes must be imaged exactly in the centre of the used slit height.

In experiments in Washington[104] the microprobe light source for qualitative analysis was imaged with a quartz lens ($f$ = 236 mm) on the slit of a Wadsworth grating spectrograph with a dispersion of 5 Å/mm in the first order. The electrodes were at a distance of 944 mm from the slit and the condenser was mounted at the midpoint of this distance on the optical bench of the spectrograph. For quantitative work the spark gap was brought to within 500 mm of the slit and a quartz cylindrical lens ($f$ = 95 mm) was placed 39 mm from the slit. The slit cover carried a lens ($f$ = 457 mm) with a platinum filter coating having transmission steps of 6, 25 and 100 per cent. The set-up for qualitative laser probe analysis yields spectra four times as intense as those formed by two condensers.

## 2.6. THE SPECTROGRAPH

Grating as well as prism spectrographs have been employed in laser micro-spectrochemical analysis. The laser probe designed by Brech and Schuch is regarded as an accessory of a 1·5 m Wadsworth concave grating spectrograph. The laser micro-analyser models LMA of VEB Carl Zeiss, Jena, were first used and shown at exhibitions with the medium quartz prism spectrograph model Q 24 already owned by many spectrochemical laboratories.

The Wadsworth spectrograph of the Jarrell-Ash company has become familiar as the main component of the comprehensive laboratory outfit 'Spec Lab'. This includes the spectrograph mounted on a multiple purpose source unit 'Standard Varisource', an electrode shaper, a photo processor for the quick development of the exposed 35 mm films, a comparator microphotometer and a calculating board for rapid quantitative spectrochemical analysis (Figs. 40 and 41). The slit mount of the spectrograph takes either a symmetrical slit of a width continuously adjustable between 5 and 400 μm or a set of six slits with different fixed widths. The optical system of the spectrograph consists of a mirror and a grating with either 1180 lines/mm or 590 lines/mm. The aperture ratio of both available models of the spectrograph, standard and wide angle, is $f/24$. The basic outfit of the 1968 wide angle model 78-090 has a grating with 590 lines/mm and covers in the first order the wavelength range from 420 to 960 nm, with a reciprocal linear dispersion of 1·08 nm/mm. The second order range is 210 to 480 nm with a dispersion of 0·54 nm/mm; and the third order range 210 to 320 nm with a dispersion of 0·36 nm/mm.

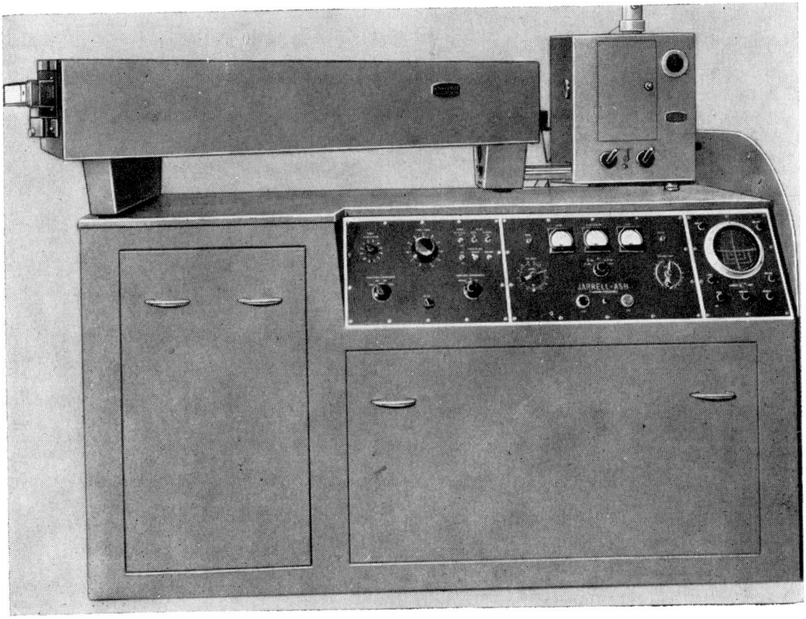


FIG. 40. 1·5 m Wadsworth grating spectrograph of Jarrell–Ash Co.

In the medium ultra-violet spectrographs of VEB Carl Zeiss, Jena, and of Hilger & Watts (now Rank Precision Industries Ltd, Analytical Division), Margate, the dispersing element is a 60° quartz Cornu prism of 63 mm and 65 mm base length respectively. Both these instruments have been employed for laser micro-spectrochemical analysis; the Zeiss Q 24 spectrograph by the authors as part of the Zeiss-LMA outfit,[13] the Hilger E 498 spectrograph by Archbold, Harper and Hughes for time-resolved spectroscopy of the microplasma produced by the laser beam focused on the surface of solid samples.[7] Valuable characteristics of these instruments are great light gathering power (maximum effective aperture ratio of the Q 24 model *f*/11·2); flat field, a great advantage in work with photographic plates when the whole spectrum is photographed on one plate, which is of special importance in micro-analysis of costly materials; and the possibility of adding a wavelength scale to the spectra on the photographic plate.

The length of the entire spectrum from 200 to 580 nm on the photographic plate of the Zeiss model Q 24 is 223 mm and that of the spectrum from 200 to 1000 nm on the plate of the Hilger model E 498 (Fig. 42a) is 220 mm. Because of the wavelength dependence of the dispersion only

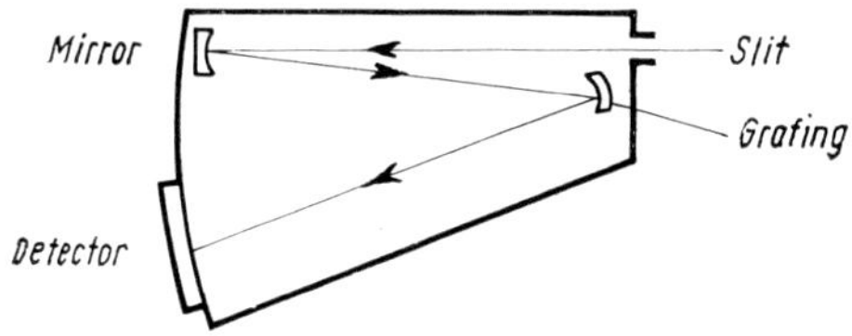


FIG. 41. Ray diagram of Wadsworth grating spectrograph.

the range up to about 460 nm is, of course, fully utilized with medium ultra-violet quartz prism spectrographs.

The Zeiss model Q 24 has a symmetrical precision slit whose height can be adjustable between 1 and 12 mm; the slit width can be read on a scalewith a precision of 1 μm. The slit mount of the present production models s fitted wiith three different length-reducing stops and two step filters for quantitative analysis. The corrected collimator and camera objectives have a diameter of 50 mm and a focal length of 540 and 500 mm respectively. The camera cassette is racked up by means of a hand wheel below the collimator.

On the photographic plate, 6 × 24 cm, a total of 45 spectra can be taken with a slit height of 1 mm which is fully adequate for laser micro-analysis. This can be increased to 75 spectra per plate by using the 9 × 24 cm plate size on the Q 24 spectrograph. The spectrum plates must, of course, be of perfect quality.

Debras-Guédon, Liodec and Vilnat[129b] at the Laboratoire de Physico-Chimie de la Société Française de Céramique, Paris, used the quartz spectrograph $Z_3$ of Jobin and Yvon, Arcueil (Seine). This instrument covers the spectrum range from 210 to 700 nm in a length of 560 mm. The collimator system consists of an aluminized concave mirror and a plane deviating mirror; the Cornu prism has a refracting angle of 65°. The camera objective consisting of three quartz lenses is a modified version of the system 'Nouvelle Zélande' ($f = 1200$ mm; $f/24$).

The Jarrell-Ash 1·5 m Wadsworth spectrograph is supplied with a rollfilm cassette; it allows use of a 50 cm film length which after exposure can be cut and removed from the light-tight cassette for development. Here the different spectra are recorded horizontally across the film, different from their vertical arrangement on the photoplates of the model Q 24 mentioned before.

a)

b)

FIG. 42. The Hilger prism spectrographs E 498(*a*) and E 478(*b*). The model E 478 is a Littrow spectrograph with interchangeable quartz and glass prisms (now superseded by new model E 742/3/4).

Runge, Minck and Bryan[113] carried out their quantitative steel analyses on a large Littrow quartz spectrograph with laser vaporization and without additional cross excitation. One of the original instruments of this design was the Zeiss model QG 55, giving a spectrum from 200 to 400 nm of about 550 mm length. A widely used interchangeable quartz and glass Littrow spectrograph is the Hilger & Watts (Rank Precision Industries Ltd.) type E 742 (Fig. 42b) with quartz and glass prism. The quartz Littrow prism has a refracting angle of 30°, a base length of 94 mm and covers the range from 191 to 800 nm in a spectrum of 670 mm length; the glass Littrow prism of 26° refracting angle and the same base length of 94 mm has a working wavelength range from 370 to 1200 nm. A great advantage of this instrument is its high dispersion in the ultra-violet; below 300 nm it is generally better than that of a grating spectrograph with a 3 m grating with 590 lines/mm. The collimator objectives of the spectrographs of this type have a focal length of about 1700 mm; the aperture ratios in the ultra-violet, however, are only of the order of *f*/35 to *f*/42. The glass optic on the other hand not only gives a three times greater dispersion in the visible but also, at 500 nm, an improved aperture ratio of about *f*/25.

To the same class of spectrographs as the Zeiss model QG 55 and the Hilger model E 742 also belong the USSR model KSA-1 and the model QL-170 of Shimadzu Seisakusho, Kyoto, Japan.

With all these prism spectrographs, as with all large grating spectrographs, the total extent of the spectrum range can only be photographed in a number of exposures of plates or films.

This also holds for the Zeiss three prism spectrograph 'Collimator 70/Camera 70'; it was available from 1964 to 1970 as a standard glass spectrograph as an addition to the model Q 24, covering the wavelength range from 370 to 1000 nm and was very useful for applications in mineralogy (Fig. 43). Its aperture ratio is *f*/13·5. The optical system consists of an achromatic doublet lens, the Försterling three prisms train and a triplet camera objective with an exceptionally large field lens of 150 mm diameter. The dispersion at 370 nm is 0·45 nm/mm and the effective resolution at 400 nm is 0·015 nm, compared with 0·077 nm for the Q 24 model. With the prism train in a medium setting the spectrum region from 380 to 670 nm was covered on a photographic plate of 18 cm length.

The standard glass spectrograph, amongst others, makes possible the laser micro-analysis of such elements as Li, Rb and Cs which have no suitable analytical lines in the working range of the Q 24 model.

Ebert plane grating spectrographs can also serve for laser micro-spectroscopy; they have, compared with concave grating instruments, the advantage of stigmatic image formation.

The laser microprobe designed by Brech and Schuch has, amongst others, also been linked with a Jarrell-Ash 3·4 m Ebert spectrograph in mineralogical studies.[74,75] The LMA-1 model of VEB Carl Zeiss, Jena,

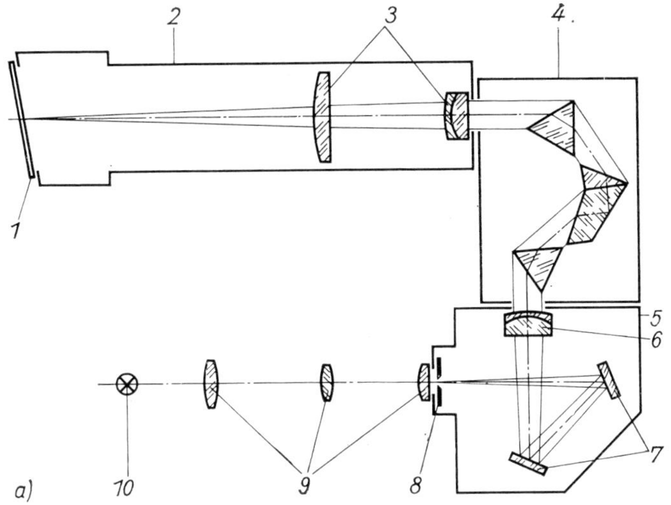

(b)

FIG. 43. Standard glass prism spectrograph of VEB Carl Zeiss, Jena.

*a*) Ray diagram: 1 Photographic plate; 2 Camera body: 3 Camera objective; 4 Prism housing; 5 Collimator housing; 6 Collimator objective; 7 Collimator deviating mirrors; 8 Entrance slit; 9 Condenser lenses; 10 Light source.

(*b*) Laser micro-analyser LMA 1 set-up with a standard glass prism spectrograph.

TABLE 6. Basic specification of spectrographs

| *Instrument model* | *Manufacturers* | *Dispersing element* | *Effective aperture ratio* | *Inverse dispersion* (nm/mm) | *Wavelength range* (nm) | *Photographic plates or films* |
|---|---|---|---|---|---|---|
| Q 24 | VEB Carl Zeiss, Jena | 60° Cornu prism | *f*/11<br>*f*/15<br>*f*/30 | 0·39 (at 200 nm)<br>1·35 (at 300 nm)<br>8·5 (at 580 nm) | 200–580 | Plates: 9 × 24 cm<br>6 × 24 cm |
| Standard glass spectrograph | VEB Carl Zeiss, Jena | Three glass prisms (Försterling sytem) | *f*/13·5 | 0·45 (at 370 nm)<br>1·7 (at 500 nm)<br>2·7 (at 600 nm) | 370–1000 | Plates: 6 × 18 cm |
| E 498 | R.P.I. Ltd, Analytical Division (Hilger and Watts), London | 60° Cornu prism | *f*/11·8 | 1·22 (at 280 nm) | 200–1000 | Plates: 4 × 10 inch<br>2 × 10 inch<br>9 × 24 cm |
| E 472/3 | R.P.I. Ltd | 30° quartz Littrow prism | *f*/24·5 | 0·41 (at 280 nm) | 191–800 | Plates: 4 × 10 inch<br>2 × 10 inch<br>9 × 24 cm |
| E 472/4 | R.P.I. Ltd | 26° glass Littrow prism | *f*/26·1 | 0·63 (at 435·8 nm) | 370–1200 | Plates: 4 × 10 inch<br>2 × 10 inch<br>9 × 24 cm |

| | | | | | | |
|---|---|---|---|---|---|---|
| STE-1 modification | LOMO, Leningrad | Two plane gratings 1200 and 600 lines/mm | $f/15$ | 1 (at 220–440 nm)<br>2 (at 240–880 nm) | 220–880 | Film: 35 mm |
| 1·5 m Wadsworth spectrograph, 'wide angle' 78-090 | Jarrel-Ash Co., Waltham, USA | Concave grating 1180 lines/mm<br>Concave grating 590 lines/mm | $f/24$ | 0·54 (1st Order)<br>0·27 (2nd Order)<br>1·08 (1st Order)<br>0·54 (2nd Order)<br>0·36 (3rd Order) | 210–480<br>210–240<br>420–960<br>210–480<br>210–320 | Film: 35 mm<br><br>Film: 35 mm |
| PGS 2 | VEB Carl Zeiss, Jena | Plane grating 650 lines/mm | $f/28{\cdot}5$ | 0·74 (1st Order)<br>0·37 (2nd Order) | 200–2800 | Plates: 9 × 24 cm |
| 3·4 m Ebert spectrograph | Jarrell-Ash Co., Waltham, USA | Plane grating<br>600 lines/mm<br>1200 lines/mm | $f/33$<br>$f/10$ | 0·51 (1st Order)<br>0·24 (2nd Order) | 180–3000<br>180–1500 | Plates: 20 inch<br>30 inch |
| 0·75 m Czerny-Turner spectrograph | Jarrell-Ash Co. | Plane grating 300 lines/mm | $f/6{\cdot}3$ | 4·4 | 200–2000 | Plates or films: 5 × 7 inch |

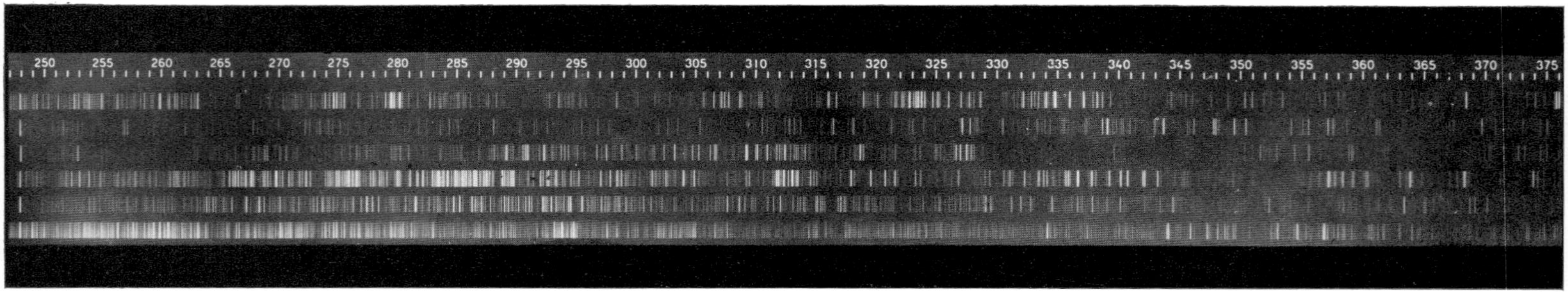


Fig. 44. Spectra of important minerals produced with LMA 1 laser micro-spectrum analyser and plane grating spectrograph PGS 2.

From top to bottom: ilmenite $FeTiO_3$; zircon $ZrSiO_4$; pucherite $BiVO_4$; crocoite $PbCrO_4$; molybdenite $MoS_2$; wolframite $(Mn, Fe)_2WO_4$. Technical details of set-up: LMA 1 spectrum analyser; ruby resonator, 2 kV, 508 $\mu$F; objective $\times$ 16/0·2; spark gap: 125 $\mu$H, 2·5 $\mu$F, 2·5 kV; grating 651 lines/mm, 1st order blaze 280 nm; single passage.

can, with a different interchangeable pedestal, be used together with their 2 m plane grating spectrograph. Because this instrument's effective aperture ratio is only *f*/28·5 several laser pulses per spectrum, when using model ZFL 750, or the higher laser energy of model ZFL 1000, are required for a spectrum taken for preliminary inspection compared with the more powerful spectrograph model Q 24 (*f*/11 or *f*/15). A wavelength scale extending to an upper limit of 700 nm in the first order is reproduced on the spectrum plate for identification within the wavelength range of interest to mineralogists and chemists (Fig. 44). The linear dispersion of the spectrograph model PGS 2 when used with an (interchangeable) grating of 650 lines/mm is 0·74 nm/mm in the first order and 0·37 nm/mm in the second order.[86] The use of the 2 m plane grating spectrograph requires some considerable experience in spectrum analysis and also some notion of the likely chemical composition of the sample because spectra for preliminary inspection, on account of the high linear dispersion, always have to be taken in sections on several plates.

Table 6 contains a summary of the specifications of the spectrographs that have already successfully been employed in combination with commercial or experimental laser excitation sources. An additional auxiliary lens system for the 3·4 m Ebert Spectrograph of the Jarrell-Ash Co. reduces the line length from 12 mm to 1·2 mm, thereby improving the aperture ratio from *f*/33 to *f*/10.

LOMO, Leningrad, developed from their model STE-1 with crossed dispersion a compact double grating spectrograph for the MSL 2 outfit (Table 6). The collimator and camera lens have a focal length of 901·5 mm and 807·2 mm, respectively. The quartz and glass prisms in front of the two gratings of model STE-1 have of course been dispensed with in the new instrument.

For certain special applications, resorting to very fast spectrographs is advisable, as for instance the Jarrell–Ash 0·75 m Czerny–Turner plane grating spectrograph with an aperture ratio of *f*/6·3. The grating of the standard version (Table 6) is interchangeable with several others having 600, 1200 or 2160 lines/mm. Such special versions have so far hardly been used for routine work and therefore comprehensive comparison data are generally not yet available.

# 3

# The Procedure of Laser Micro-Spectrochemical Analysis

## 3.1. FORMS OF SPECTROCHEMICAL ANALYSIS WITH SAMPLE VAPORIZATION BY LASER RADIATION

The combination of a solid-state laser and a special microscope aligned on an optical bench takes the place of the usual discharge stand and source unit (for spark, continuous or intermittent arc generation) of conventional spectrochemical analysis. In most cases commercial spectrographs are used which as a rule are suitable without any modification for conventional analysis of quantities of the gram and milligram order as well as for laser micro-analysis of micrograms.

There are basically two methods of laser micro-analysis, each with two variants (see Table 8). In method I the microplasma produced by the laser radiation is imaged, preferably be reflecting optics, on the slit of the spectrograph. In method II an auxiliary spark gap is placed between the sample and the object and the spark discharge serves as spectroscopic source ['cross excitation' - in the USSR called 'Double Supply' (Sswentizki[63a])]. A voltage of several thousand volts is applied to the two suitably mounted and insulated electrodes.

In both methods the laser provides the energy for the sample vaporization and its radiation is concentrated by means of a microscope objective on a region of the sample first selected visually in an examination under the microscope. A portion of the coherent red or infra-red radiation is absorbed by the sample.

The material irradiated by the laser radiation is vaporized as result of the exceedingly high energy density and temperature. A microplasma is produced close to the sample surface and often red-hot particles of material are ejected from the impact crater.

In photographs of the laser impact taken with an ordinary miniature camera, which can be fixed on an optical bench carriage, exactly rectilinear flight paths described by the spattered particles under the influence of the high kinetic energy are observed – especially so with inorganic chromium compounds like chromium oxide minerals, for example $FeCr_2O_4$ or chromium steels. The particles are often reflected by the front protecting plate of the objective and, on rebounding, by the sample surface (Fig. 45).

TABLE 7. Scope and limitation of instrumentation for laser micro-spectrochemical analysis (according to Laqua, published by permission of the author)

| | *Q-switched laser* | | | | *Non-switched laser* | | | |
|---|---|---|---|---|---|---|---|---|
| | *With cross excitation* | | *Without cross excitation* | | *With cross excitation* | | *Without cross excitation* | |
| | *Photographic plate* | *Photo-multiplier* | *Photographic plate* | *Photo-multiplier* | *Photographic plate* | *Photo-multiplier* | *Photographic plate* | *Photo-multiplier* |
| 2·2 m grating spectrograph *f*/32 | − | + | | + | − | + | − | − |
| Q 24 prism spectrograph *f*/11 | + | | − | | + | | − | |
| 0·45 m Czerny-Turner grating spectrograph *f*/5 | + | | + | | + | | − | |

\+ Good results − Not feasible.

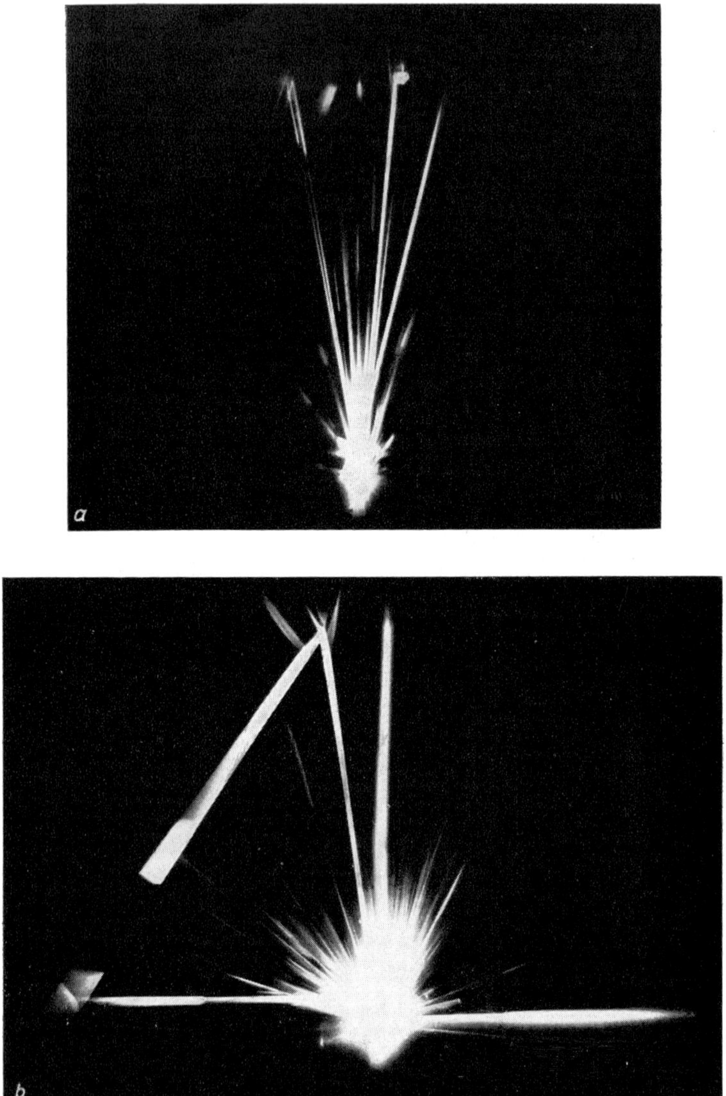


FIG. 45. Effect of wavelength and laser output energy on microplasma dimensions and kinetic energy of ejected sample particles.

Sample substance: $FeCr_2O_4$.

(*a*) Wavelength 694·3 nm; Output 0·03 J.

(*b*) Wavelength 1060 nm; Output 0·4 J.

The rotation of unsymmetrically heated chromite partices manifests itself in a rectilinear helical luminous track (Fig. 47). Frequently ejected particles burst in flight between the sample surface and the microscope objective (Fig. 46).

The various possibilities of laser spot analysis were investigated at the Institute for Spectrochemistry, Dortmund, and the performances of the following instruments were compared: a 2·2 m grating spectrograph, the Zeiss quartz spectrograph model Q 24, and a 0·45 m grating spectrograph of aperture ratio *f*/5 built at the Institute with a Bausch & Lomb grating having 2160 lines/mm. The results were reported by Laqua at the Jena Meeting 'Optics and Spectroscopy of all wavelength regions' in 1965 and are summarized in Table 7. They show that medium prism spectrographs,

FIG. 46. Steel particles (X22CrNi17) disintegrating in bursts under the effect of focused laser radiation (neodymium glass resonator; wavelength 1060 nm; output energy 0·5 J).

and also concave grating spectrographs of short focal lengths and great light gathering powers, should preferably be used only for recording the spectra on conventional photographic plates. On the other hand, photoelectric spectrometers offer greater detector sensitivity but, because of the restricted number of photomultipliers, do not allow comprehensive spectra to be taken for preliminary inspection. It emerges furthermore

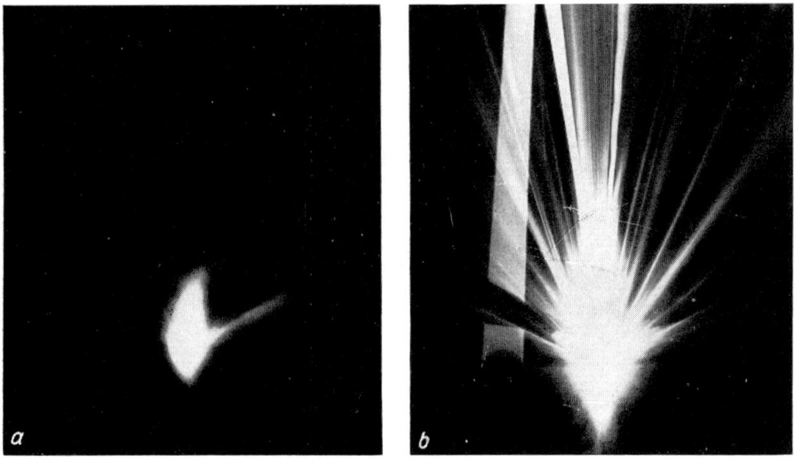


FIG. 47. Microplasma generated on polished section of chromite.

(*a*) Infra-red laser radiation ($\lambda$ = 1060 nm) focused with reflecting microscope objective $\times$40/0·5, $\infty$/0, focal length 6·3 mm.

(*b*) Infra-red laser radiation ($\lambda$ = 1060 nm) focused with refracting microscope objective $\times$16/0·2, $\infty$/2, focal length 16 mm.

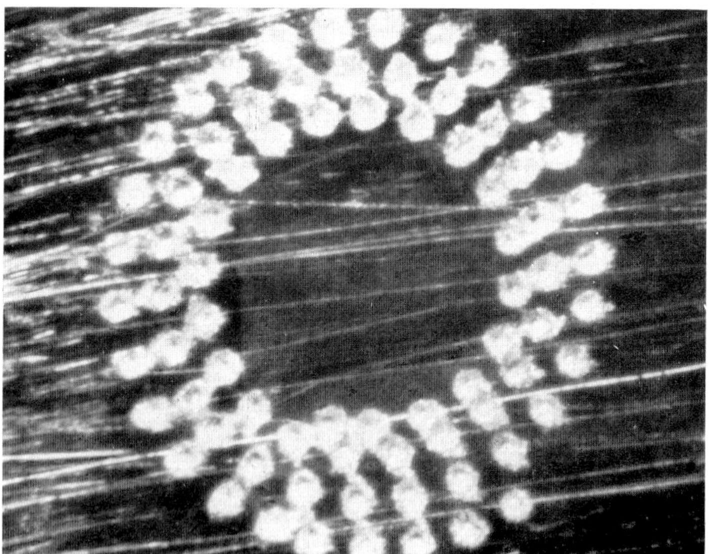

FIG. 48. Scanning method of laser micro-spectrochemical analysis for obtaining an average analysis (laser macro-analysis).

that a laser micro-analysis outfit with an ultra-violet spectrograph definitely should provide cross excitation, which means including the adjustable electrode holders and mains supply unit for it.

Q-switched lasers have the advantage over non-switched solid-state lasers that they sometimes allow work without cross excitation. (Problems of spectrum interpretation are discussed in § 3.4.3.3.)

There are two further methods, III and IV, of laser spectrum analysis to be classified as methods of laser macro-spectrochemical analysis. They were developed from micro-analysis and consist in firing a series of laser shots across the sample surface in a regular manner and recording the results in a single spectrum (Fig. 48).

Only method IIIb promises success because auxiliary spark excitation is problematic (great spectrum density, necessity of automatic electrode setting control), but is necessary for work with a non-switched laser. Two further methods, V and VI, are also classified as laser macro-analysis: here the laser radiation is focused not with a microscope, but with a single glass lens of short focal length. Craters of comparatively large diameters, namely several hundred micrometres or more, are produced. This book

TABLE 8. Methods of emission spectrochemical analysis with solid-state lasers

Method I: Micro-analysis of the plasma on the sample surface

(*a*) with non-switched pulsed laser and with prism or grating spectrograph
(*b*) with Q-switched laser and with prism or grating spectrograph

Method II: Micro-analysis with auxiliary spark gap between sample and objective

(*a*) as in method I
(*b*) as in method I

Method III: Macro-analysis by summation of micro-analyses (scanning procedure, pulsed laser with high repetition rate; mainly without spark excitation)

(*a*) with non-switched laser: not promising
(*b*) with Q-switched laser: grating spectrograph or spectrometer

Method IV: Macro-analysis by summation of micro-analyses with spark excitation (problematic)

(*a*) as in method III
(*b*) as in method III

Method V: Macro-analysis, focusing laser radiation with single, short-focal-length lens, without spark excitation

(*a*) with non-switched pulsed laser and with prism or grating spectrograph. Detector: photographic plate, film, photomultiplier.
(*b*) with Q-switched laser and with prism or grating spectrograph. Detector: photographic plate, film, photomultiplier

Method VI: Macro-analysis as method V, but with spark excitation

(*a*) as in method V
(*b*) as in method V.

in the main deals with methods I and II; the methods of macro-analysis are only briefly referred to when necessary.

In France Debras-Guédon and Liodec[27,28] used a non-switched ruby laser without cross excitation for the analysis of different materials. The first experiments did not yield any useful results; after more work optimum working conditions were established. They differ considerably in parts from the usual instrumental arrangements: the spectrograph, not described in detail ['... avec un spectrographe de bonnes dispersion et luminosité...' (with a spectrograph of high dispersion and light gathering power)], has a horizontal entrance slit. The sample surface is fixed at right angles to the supporting table. A spherical mirror is placed behind the irradiated sample spot for concentrating more energy into the image of the microplasma which emits in all directions. A special device for sample and mirror alignment had to be constructed. Although a crater diameter of about 0·1 mm was obtained the spectra recorded on rapid photographic plates did not contain any lines of the elements present in the sample, namely Ag, B, Be, Bi, C, Cd, Co, Hg, Li, P and Sn (Method V(a)).

According to the investigations of Runge, Bonfiglio and Bryan[112] the Method V(b) seems to be of exceptional importance for the analysis of samples in the molten state, still in the furnace. These workers used a Q-switched ruby laser with pulse powers of 3 to 5 MW and durations of 15–20 ns, that is energies of about 50 to 75 mJ; the coherent radiation from the laser was focused on the surface of the melt with a lens of 15 mm focal length. The plasma was imaged on the slit of a quartz Littrow spectrograph by means of a quartz condenser and a deviating mirror. Three laser pulses were used for each spectrum and, according to the report, a sample volume of about $10^{-6}$ cm$^3$ was vaporized by each pulse. The spectra were photographed with Kodak plates 103-0. The craters in metal surfaces had diameters of about 0·5 mm and the plasma cloud a diameter of about 3 mm. The plasma was whitish-blue against the yellowish-red background formed by the melt. The strong background continuum did not interfere with the evaluation of line pairs of some main constituents of steel melts. Rectilinear working graphs were obtained for Ni 341·48 nm/Fe 376·55 nm and Cr 396·37 nm/Fe 376·55 nm with solid steel standard samples as well as in the examination of the material afterwards melted in the induction furnace.

The problem of such investigations, so far unsolved, but most important for practical applications, is the determination of additional elements present in the metal melt and in the steel in low concentrations.

Runge, Minck and Bryan[113] of the Scientific Laboratory of the Ford Motor Company, Dearborn, Michigan, USA, used a 3 MW Q-switched ruby laser for quantitative steel analysis. This is a typical macro-analysis according to Method VI(b), with crater diameters of about 0·4 mm. Each sample was irradiated with four laser pulses to produce spectrograms suitable for photometric evaluation.

*Examination of the microplasma produced by concentrated laser radiation.*

The methods of time-resolved spectroscopy and the instruments for high-speed photography can be employed for the analysis of the microplasma produced by laser radiation.

Scott and Strasheim[118] found the following connections between the Laser working mode, the speed of plasma expansion and the crater dimensions:

| *Laser working mode* | *Plasma expansion speed* | *Crater* Diameter (μm) | *Crater* Depth (μm) |
|---|---|---|---|
| Non-switched | $10^4$ cm s$^{-1}$ | 200 | 1150 |
| | | (Intense particle spattering) | |
| Semi-Q-switched: rotating prism (3000 r.p.m.) | $10^5$ cm s$^{-1}$ | 250 | 100 |
| Q-switched single pulse | $10^6$ cm s$^{-1}$ | 250 | 10 |

The spectra obtained without Q-switching and without cross excitation show marked self-absorption. Typical for semi-Q-switching are doubly ionized spectrum lines which can only just be detected or not at all in a plasma produced by a single-pulse Q-switched laser. Single pulse working leads to narrow line width, but also to strong background radiation in the wavelength range from 355 to 458 nm.

Panteleev and Jankowsky[96] reported expansion velocities of $2 \times 10^6$ cm s$^{-1}$ at a plasma temperature of $10^5$ °K, with a pulse energy of 0·5 J and a pulse duration of 50 ns. The peak radiation density is reached up to 10 μs after the start of the laser impact on the sample surface; over a period of a further 30 μs, a second, weaker phase is observed. 100 μs after the interaction of radiation and mass the flare has ceased and, with use of a focusing lens of 120 mm focal length, a crater of 1 mm diameter and 1 mm depth remains.

Workers in the USSR predict a future of great importance for single-pulse working for determining surface impurities.

In 1964 Archbold, Harper and Hughes used a 1 MW Q-switched ruby laser to obtain time-resolved spectra of the microplasma produced on samples of Al, Cu, Zn, Ge, Sn, W, Au, Pb, borosilicate and crown glass and ruby. Their laser had a ruby resonator doped with 0·05 per cent $Cr_2O_3$, fitted together with a collinear xenon flash tube in a spherical reflector. The Q-switch reflector is a block of special flint glass and rotates with a speed of 100 r.p.s. about an axis parallel to its polished plane-parallel reflecting surfaces and to the surface of the resonator. After emerging from the glass block and after reflection by a semi-reflecting flat glass plate, part of the laser radiation falls on the cathode of a photocell connected to an oscillograph for recording intensity against time curves. The other part of the coherent monochromatic radiation passes through the glass plate, a deviating prism and a lens of short focal length, namely 12 mm, and is focused by the latter on the sample surface. A system of three front-aluminized mirrors images the plasma on the slit of a Hilger quartz spectrograph. One of the mirrors is mounted on the axis of the motor driving the Q-switch reflector and produces an image sweep speed of 0·47 mm $\mu s^{-1}$. The rotating mirror also directs a white light beam on to a photomultiplier and thereby provides a suitable trigger pulse for the flash tube in the laser head.

Interpretation of the emission spectra obtained at the National Physical Laboratory, Teddington, England,[7] revealed that a highly ionized microplasma is formed near the irradiated sample surface. During the lifetime of the laser pulse (0·1 to 0·5 μs) a strictly continuous emission down to a wavelength limit of 230 nm was observed. In addition to the strongest spectrum lines of the neutral atoms and the level II lines of the singly ionized atoms, spectrum lines of the third ionization level (Al III, Sn III, Pb III, Ge III) and even short-life Sn IV and Ge IV lines could be determined. These lines of the higher ionization levels are only emitted for relatively short periods; therefore the plasma is believed to cool very quickly. The lines of the first and second spectra, on the other hand, can also be observed in between the individual laser pulses, and the atomic lines spectra remain identifiable for a comparatively long time period of several microseconds after the end of the laser action.

Apart from normal self-absorption of lines of the first spectrum, asymmetrical self-absorption has been observed, especially with Sn I. There is no continuum in the centres and therefore it can be concluded that a fairly large number of neutral atoms could absorb radiation between the sample and the spectrograph about $10^{-7}$ s after the start of the laser impact, that is before the target area on the sample surface can be heated to incandescence. Although apparently highly transparent for the long wavelength laser radiation, this vapour seems to be of great importance for the heating mechanism because the radiation density is of the order of $10^{10}$ W/cm$^2$. The order of magnitude of the temperature in the emitting sample region is of interest. According to Archbold, Harper and Hughes[7] it is about 5000 °K, the emitting surface being assumed to be a black body radiator. The complete absence of band spectra and of oxygen and nitrogen lines is notable in view of the fact that the samples were irradiated in air. The Ge IV and Sn IV emission indicates an electron temperature of about 20000 °K.

Kaporski, Ossipov and Sswentizki[63a] found a connection between laser spike series and vaporization of discrete masses of material (single eruptions). Conclusions about density and temperature differences within the plasma flare are drawn from the nature of recorded spectrum lines

TABLE 9. Development of emission spectrum analysis with solid-state laser source for sample evaporization 1962–1966

| *Stage of Development* | *Author* | *Year* | *Optic between resonator and sample* | *Spectrograph* |
|---|---|---|---|---|
| Spot analysis, sample vaporization by Q-switched ruby laser, 0·2 J, 20 ns | Brech *et al.* | 1962/3 | Achromat, ×20/0·25 | 1·5 m Wadsworth concave grating system |
| Macro-analysis (vaporization and spectrum excitation) of steel samples with Q-switched ruby laser, MW | Runge, Minck, Bryan | 1963 | Single lens | Quartz Littrow spectrograph |
| Macro-analysis with non-switched ruby laser, 100 J | Liodec, Debras-Guédon | 1963 | Single lens | Medium quartz spectrograph $Z_3$ |
| Spot analysis | Laqua, Hagenah | 1964 | Refracting objectives | Medium quartz spectrograph Zeiss Q 24, grating spectrographs |
| Micro-analysis (opaque and transparent samples) with non-switched solid-state laser, 0·2 J (wavelengths 1060 and 694·3 nm) | Moenke | 1964 | Reflecting objective ×40/0·5<br>Flat-field achromat ×16/0·2 | Medium quartz spectrograph Zeiss Q 24<br>2m Ebert plane grating spectrograph |
| Macro-analysis with and without cross excitation, non-switched solid-state laser, 10 J | Karyakin, Achmanova, Kaigorodov | 1965 | Single lens | Medium quartz spectrograph ISP-22 |
| Area analysis with Q-switched pulsed ruby laser, 0·1 J, high repetition rate (⩾1 Hz) | Felske, Hagenah, Laqua | 1965/6 | Monochromat $f/1{\cdot}5$, $f = 15$ mm | Medium quartz spectrograph Zeiss Q 24 |
| Analysis of main constituents of melts in furnace, Q-switched ruby laser, MW | Runge, Bonfiglio, Bryan | 1965/6 | Single lens, $f = 15$ mm | Quartz Littrow spectrograph |

(mainly atomic lines and lines of singly ionized atoms, but also Cu III, Al III and C III lines) and their spatial distribution within the flare. The occurrence of numerous lines with self-absorption is pointed out as typical.

Kaporski and Schlepkova[63a] estimated the temperature of the microplasma in air and in vacuum from Cu I, Mg II and Al III lines.

Bakanowitsch, Gretschichin and Minko[63a] described the formation of a shock wave in the ultrasonic microplasma beam (about $10^4$ °K) and determined axial and radial temperature and concentration distribution of charged particles. They do not believe that there is complete thermodynamic equilibrium although excitation by a pulsed laser is of a thermal nature.

### *High-temperature flame spectroscopy*

Stoicheff[124] describes as high-temperature flame spectroscopy a variant of laser spectrochemical analysis Method V that within certain limits permits the determination of chemical elements present in matrix substances as impurities.

If laser radiation is focused on the surface of a material rich in carbon or boron, red hot particles are ejected from the solid or powder sample and in case of carbon a bluish flame occurs. If the radiation emitted in air or in a vacuum is analysed with a medium prism spectrograph with photographic recording three main types of spectra will be found, namely: band spectra of $C_2$, CN and $BO_2$, intense continuous spectra in certain wavelength regions and line spectra of minor constituents.

Stoicheff expects new information on spectra of ions, free radicals and molecules from research with such instrumentation; in 1962 he had already recommended kinetic studies of the flares making use of time-resolved spectra.

In fluorescence spectroscopy with laser excitation the time and space coherent radiation is not used to vaporize small fractions of the sample, but is focused so that it irradiates the largest possible sample volume (see § 4.4).

## 3.2. CHOICE OF LASER AND OTHER OUTFIT UNITS

In general, problems of producing the laser radiation, even in applications of the laser to processing of materials or in spectrum analysis, are the concern of physical research institutions and not of spectrochemical laboratories. The laser user in such laboratories will only rarely dismantle the unit unless it be for servicing (for example cleaning or replacing the resonator) or for adapting it to different working methods. He should, however, be aware of possible differences between available units with regard to geometrical layout and energy characteristics, and should know where fundamental research in this field has already been carried out.

According to Lenz[66b] low $Cr^{3+}$ doping of ruby resonators necessitates high pumping energies; for strongly doped resonators on the other hand additional cooling to low temperatures and reductions of their dimensions are necessary.

Features that may affect the laser performance are for instance: the time, space and spectral distribution of the pumping radiation; the form of the xenon tube (helical or straight); the type of reflector for the pumping source; the shape of the resonator crystal, its heterogeneities, its axis orientation (preferably 0°, 60° or 90°), the coating of its end surfaces and its temperature. In Europe such problems are being investigated at, amongst others, the Institute for Optics and Spectroscopy of the German Academy of Sciences, Berlin, and the Institute for Spectrochemistry and Applied Spectroscopy, Dortmund, G.F.R. (Felske, Hagenah, Laqua[32, 33]).

To the question of choice between pulsed lasers with or without Q-switch the following can be said: a non-switched laser is adequate for the analysis of inorganic anisodesmic and of many mesodesmic materials. Crystals with isodesmic structure, especially infra-red transmitting compounds such as the alkali halides, should be irradiated with a Q-switched laser, and likewise samples which are to be analysed in relatively thin slices of a few micrometres thickness. A resonator without a Q-switch usually gives deeper craters than a giant pulse laser. For problems of geology and mineralogy an outfit with a simple pulsed solid-state laser is useful. If only sulphide ores are to be analysed a laser output of about 0·1 J is adequate; otherwise an instrument with output energy of 0·5 to 1 J should probably best be chosen.

It also has been found that under certain conditions the reproducibility of quantitative laser micro-spectroscopy is better in single pulse working. On the other hand multiple spike operation (laser without Q-switch) offers greater possibility of detection, often also evaporation of a larger sample volume.[103]

The method of Q-switching too is of importance for the crater dimensions. At the Institute for Spectrochemistry, Dortmund, a ruby laser was used first without Q-switch (2 kW per pulse), then Q-switched with a non-linear filter (200 kW per pulse) and finally Q-switched with a fast rotating interferometer output mirror (20 kW per pulse); the effect of these changes in the mode of operation was an increase of the crater diameter for iron samples from 96 μm to 108 μm and 126 μm, and for copper samples from 60 μm to 96 μm and 156 μm. The quantity of vaporized substance did not, however, vary in the same order because the crater depth was affected in a different way: it was smallest in Q-switching with the non-linear filter, namely 60 μm for iron and 200 μm for copper; the corresponding values in work without the Q-switch were 400 μm for iron and 200 μm for copper, and finally in Q-switching with the rotating interferometer mirror 130 μm and 240 μm for iron and copper, respectively. Here is a summary of the results:

| *Crater* | *Sample* | *Laser* *Without Q-switch* | *Q-switched with non-linear filter* | *Q-switched with fast rotating interferometer mirror* |
|---|---|---|---|---|
| *Diameter* | Iron | 96 | 108 | 126 |
| μm | Copper | 60 | 96 | 156 |
| *Depth* | Iron | 400 | 60 | 130 |
| μm | Copper | 200 | 200 | 240 |

All these experiments, reported by Laqua at the Conference on 'Optics and Spectroscopy of all Wavelength Regions' held at Jena in 1965, were carried out with the same output energy of 0·3 J.

A liquid cell for passive Q-switching can be fitted on a dovetail ring to the laser head of the Zeiss LMA 1 outfit (Fig. 49). The cell, of 5 mm path length and with two filling tubes, is combined with a highly reflecting fused silica prism to form the end component of the resonator system. The resonator itself is a ruby rod with plane-parallel polished end surfaces, or a rod of neodymium glass* suitable for Q-switching. The output mirror can be of BK 7 glass with a dielectric coating; it is fixed in the resonator head in a mount on the tube that carries the deviating prism for the microscope. Crypto-cyanine as Q-switch liquid should be used in a concentration of about 0·5 mg/l methanol. The transmission of this solution in the 5 mm pathlength cell can be measured on the Zeiss photoelectric photometer model G II with a metal coating interference filter SIF 693 in the light path, the transmission of the cell filled with the pure solvent being taken as 100 per cent. Any number of spikes from 1 to 100 for each flash tube pulse can be obtained through varying the concentration of the Q-switch solution.

* Normal borate glasses, such as the Kodak laser glass ND-10a with 1 to 3 per cent $Nd_2O_3$ [$n_{1 \cdot 06\,\mu m} = 1{\cdot}644$; threshold value for rods 0·125 × 3 inch (3·2 × 76 mm) about 12 J] are not suitable for giant pulse working. For these glasses the permissible energy output is about 3 J/cm$^3$. In 1964/65 the Kodak Nd-11 silicate glass ($n_{1 \cdot 06\,\mu m} = 1{\cdot}5801$) was especially developed for Q-switching and the borate glass YB-20 with 1, 3 or 6 per cent $Yb_2O_3$ ($n_{1 \cdot 02\,\mu m} = 1{\cdot}644$) for work in the megawatt range (laser wavelength 1020 nm).

FIG. 49. Dye cell and resonator set-up for passive Q-switching of laser in LMA 1 outfit (safety door opened).

## 3.3. FOCAL SPOT SIZE AND EXTENT OF IRRADIATION EFFECT. CHOICE AND UTILIZATION OF MICROSCOPE OBJECTIVES, LASER RESONATORS AND OPTICAL ATTACHMENTS

Even visual examination of the region of impact of focused red and infra-red laser radiation will make manifest the sometimes considerable difference between theoretical focal spot size, the size of the focal spot actually produced (inner crater diameter) and the dimensions of the volume affected by the irradiation (extent of shift of material measurable with microscope). Solid materials of the most diverse nature show, after irradiation by a focused coherent laser beam, an almost circular crater, often surrounded by a fusion ridge.

For practical work in laser micro-spectroscopy it is important to know within which limits and by what means the crater dimensions in the $x$–$y$ plane and the depth in the $z$ direction can be varied and how the extent of the external irradiation effects can be confined.

The crater diameter depends on electrical parameters of the energy source (laser mains supply unit), on optical characteristics of the laser microscope and the laser head, and on certain properties of the sample

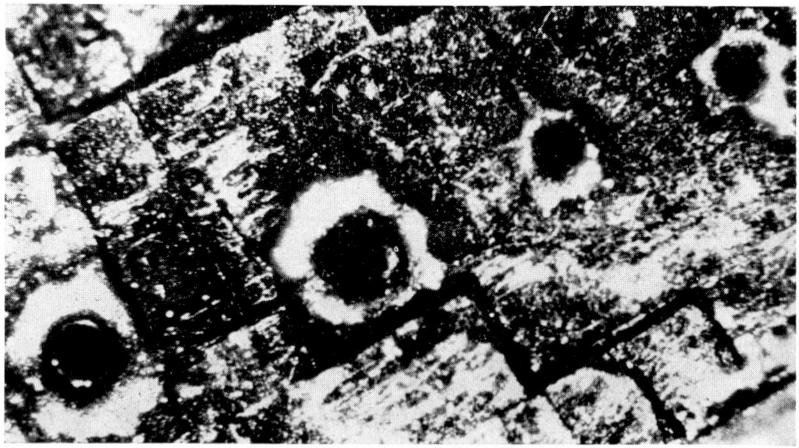

FIG. 50. Variations of crater diameter with type of microscope objective. Effects of laser radiation of wavelength 1060 nm.

*Left:* Two craters obtained by focusing with flat-field achromat ×16/0·2, ∞/2.
*Right:* Two craters obtained by focusing with reflecting objective ×40/0·5, ∞/2.
Sample material: PbS.

material such as thermal conductivity, reflectivity, light transmission, surface quality, etc.

On these sample properties also depend, in some complicated way, the focal spot size as well as, partly even to a greater degree, the extent of the irradiation effects. These influences however are nearly almost the same for different objectives (Figs. 50 and 51) and different resonators that may be used, hence a new method of material identification can be introduced which is of significance particularly for mineralogy and its special branch

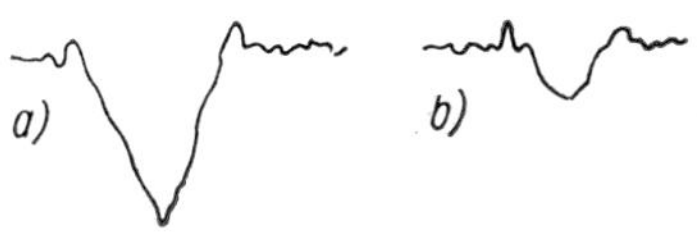


FIG. 51. Cross-sections of craters in chromium steel (vertical magnification ×1000, horizontal magnification ×100), measured with stylus probe ME 10 of VEB Carl Zeiss, Jena.

Craters produced with the laser micro-spectrum analyser LMA fitted with the laser ZFL 750 ($Nd^{3+}$ glass resonator) and (*a*) objective ×16/0·2, focal length 16 mm, (*b*) objective ×40/0·5, focal length 6·3 mm.

of petrography. Thus the very marked fusion ridge (Fig. 50) is as characteristic for the family of the lead sulphides as the always streak-like spatter trajectories for pyrite and the crater ridge with small fusion beads (Fig. 52) for erythrite.

The laser irradiation effects on pure metals and their alloys depend in a similar manner on the nature of the substance: in magnesium rather irregular fusion beads are segregated; aluminium gives small crater ridges, lead hemispherical craters; and typical for steel are a ring-shaped inner crater structure and asymmetrical fusion ridges. Sometimes the laser craters follow the outlines of inclusions.

For the laser microprobe (1964) of the Jarrell-Ash Co. the crater diameter is stated to be 50 μm. Maxwell[74] occasionally obtained such values with this instrument. It has, however, to be noted that, because of the dependence of the diameter on the material properties, such general statements have no great significance.

A Q-switched laser of the Lear Siegler Inc. has been introduced as a component of the Jarrell-Ash laser microprobe. The divergence of the light beam from the ruby laser of the LS 2 model is given in the catalogue as less than twenty minutes of arc.[51]

The crater diameter is usually considerably larger than the calculated diameter of the focal spot. The value of the latter, arrived at on the assumption of a theoretical parallel light beam, of the performance figures of

FIG. 52. Four craters of 80 μm diameter resulting from laser radiation impact on a needle crystal, 1·2 mm wide, of the cobalt arsenate erythrite $Co_3[AsO_4]_2 \cdot 8H_2O$.

a ruby laser operating at room temperature, and of the specification of the Zeiss reflecting objective of ×40/0·50, $f = 6{\cdot}3$ mm, is

$$x = \frac{\lambda}{A} = 1{\cdot}4\ \mu\text{m}$$

For the infra-red wavelength (1·06 μm) of the neodymium resonator this value increases to 2·1 μm.

If, however, the resonator diameter is not, by optical means, given an effective value $d_e = d/n$ equal to that of the entrance pupil of a precision objective, crater diameters on the sample surface of at least 40 μm result. These findings are of great importance for micro-spectrochemical analysis and confirm the small divergence of the radiation emitted by a laser resonator. The divergence $\sigma$ of the beam from a good neodymium glass rod is an order of magnitude smaller than that of a ruby rod beam – the corundum lattice is adversely affected by growth irregularities. Changing from a ruby to a neodymium glass resonator, however, will not alter the vaporized sample mass if the same microscope objective is used. A divergence diaphragm system (optical adaptor, see § 2.3.2) generally provides for both resonator types optimum adjustment with regard to the entrance pupil of the objective and at the same time a reduction of the crater diameter to about 10 μm.

The focal length of a microscope objective is of considerable practical importance because of the relationship $x = 2\,f\sigma$. In 1965 the authors introduced into laser micro-spectroscopy a lens system designed by Riesenberg; it is a flat-field achromat ×16/0·20 without cemented components and has a long working distance, namely 20 mm without and 14 mm with a protecting plate. The focal length of 15·6 mm is the best possible for vaporizing a large sample mass with crater diameters of 80–250 μm; and, in work without Q-switching, the total energy of 0·4 J effectively concentrated on steel and silicon samples is likewise an optimum value.

The effects of the divergence difference of the two most frequently used resonator materials are compensated in this objective too by the wavelength dependence of the theoretical focal spot diameter. The focal length of the objective and its transmission, which compares favourably with that of a reflecting objective with central vignetting, allow the vaporization of fairly large quantities of sample material, adequate also for determining trace element concentrations.

If crater diameters of less than 100 μm are normally required, objectives with shorter focal lengths have to be used. The reflecting objective ×40/0·50, ∞/0, of focal length 6·3 mm has proved to be satisfactory. Its diameter is equal to that of the entrance pupil of the flat-field achromat ×16/0·20; hence the best possible correspondence between the optical systems of the laser resonators and the two special objectives is assured. This fact is especially important for a reflecting objective with its central aperture restriction in the interest of avoiding unnecessary large energy losses. The Zeiss resonators are so designed that the effective emitting surface areas equal the area of the entrance pupil of the focusing objectives.

The required minimum total energy for the analysis of silicates with the reflecting objective ×40/0·50 and laser operation without Q-switching was found to be about 0·1 J. An energy of twice this value gives better qualitative results, as for instance in the analysis of the scandium silicate thortveitite $Sc_2Si_2O_7$ (50 μm crater diameter without optical adaptor). An energy of 0·03 J emitted by the optical system of the outfit is sufficient for vaporizing sulphides. A value of 0·02 J has been established as minimum necessary energy for determining iron in steel.

Beyer[129a] used immersion objectives ×90/1·25 in an irradiation unit with a pulsed ruby laser and a divergence stop on biological samples and thereby could reduce the irradiated area to a few micrometres.

The diagram of an optical system for directing and focusing the laser light beam developed by French scientists[5] is shown in Fig. 53. The transparent sample rests on a dielectric reflector *M*. The microscope is provided with an illuminator for work with reflected light. The sample is centred by viewing its secondary image formed by autocollimation by the exit surface *m* of the resonator *L*. The light intensity can be controlled by means of the polarizers $P_1$ and $P_2$ and of a quarter-wave plate $Q$ which has to be removed during the irradiation by the laser. An interference contrast method, originally introduced by Nomarski for metallographic investigations, is used for viewing. The focal spot can be aligned by making use of

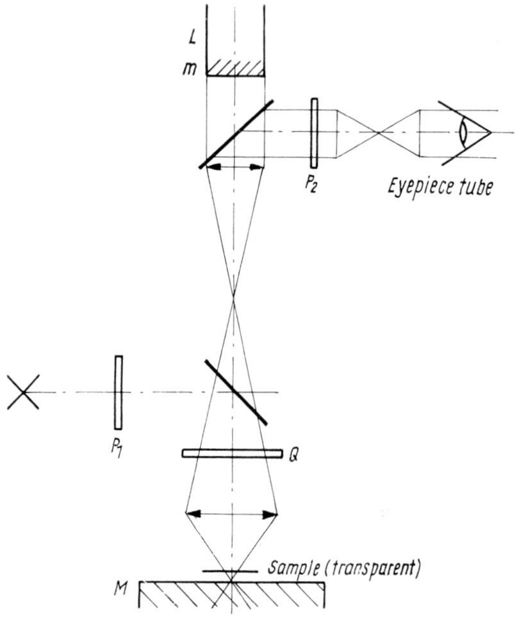


FIG. 53. Diagram of arrangement[5] for focusing and localizing red laser radiation on target details in transparent samples.

a faint image reflected by the laser; during this adjustment the quarter-wave plate is moved out of the light beam. This outfit also includes a television unit made by CSF, Corbeville, which permits direct observation of the sample during the impact of the focused coherent laser radiation. Such a facility does not serve a useful purpose in routine laser spectrum analysis, either on the micro- or the macro-scale. But the provision of an interchangeable telescope system for a laser spectrum analyser is desirable for bringing the laser beam to an intermediate focus in the plane of a divergence diaphragm; in this way the best possible adaptation of the beam width to the entrance pupil of the objective can be achieved and the crater diameter can be reduced to values between 40 and about 10 μm.

Also among the optical auxiliary units of laser spectrum analysers are deviators. They are required when the optical axes of the resonator and microscope are not collinear. The deviator of the laser microprobe designed by Brech and Schuch is essentially a 90° prism; the radiation beam emerges from the resonator almost parallel to the optical bench of the spectrograph. If a small solid-state laser with an output energy up to about 0·2 J is used for the micro-analysis, such an arrangement is better avoided by fixing the resonator housing with a support rod on the microscope tube.[13, 128]

The Zeiss model LMA 1 also includes a 90° deviating prism because a laser with adequate output energy, namely the model ZFL, is available.

If the laser system has two separate external reflectors, as many USSR designs and also the instrument of the Institute for Spectrochemistry, Dortmund, G.F.R., there are further possibilities of effectively reducing the beam divergence: a telescope system with a diaphragm can be introduced between the uncoated cylindrical ruby rod and one of the external mirrors (two alternatives); or a second Fabry-Pérot interferometer, with its plates inclined to those of the main laser interferometer, can be placed between the fully reflecting external mirror and the laser rod (Fig. 2k) – such an arrangement was described by Felske, Hagenah, Klocke and Laqua in 1965. The simplest, but least efficient solution with regard to energy, namely a diaphragm in any position between the output mirror and the objective, is usually not favoured, not even for systems with external mirrors.

The provision of additional optical systems between the laser resonator and the microscope objective at first sight seems to be very promising, but it leads to a considerable loss of output energy: an output energy of 0·5 J at the partially reflecting resonator end is reduced to 0·03 J with following transmission through the objective.[94]

TABLE 10. Dependence of crater dimensions on laser output

Material: sintered uranium oxide
Laser radiation: wavelength 1060 nm, without Q-switch, focused on sample with objective of 32 mm focal length[128]

| *Crater diameter* (μm) | *Crater depth* (μm) | *Laser output (without Q-switching)* (J) |
|---|---|---|
| 62 | 75 | 0·21 |
| 94 | 90 | 0·23 |
| 125 | 110 | 0·25 |
| 156 | 145 | 0·29 |

A method of altering the crater dimensions frequently employed in micro-spectroscopy consists in varying the flash tube voltage. Even the second decimal figure of the laser output value is significant as Table 10 shows.[128]

When in laser irradiation of steel type 40 NiCrMo 15 with the LMA 1 model the voltage of the pumping capacitor bank is lowered from the maximum value of 2·0 kV to a value of 1·5 kV, then according to Quillfeldt the ratio of crater diameter to crater depth changes from 1·7 to 10; the actual dimensions of crater diameter and crater depth are reduced, without use of the optical adaptor diaphragm, from 190 to 40 μm and from 110 to 4 μm, respectively.

Continuously increasing the pumping energy in irradiation of steel samples leads first, that is for voltages up to 1·8 kV, to a rapid lengthening of the crater diameter; if for a resonator without Q-switch, considered here, the voltage is raised further from 1·8 to 2·0 kV, the inner crater diameter remains almost unchanged, but the crater depth continuously increases from about 50 to 110 μm.

Of very great importance for micro-spectrochemical analysis is the relationship between vaporized crater volume $V$ and the intensity $I$ of the spectrum lines of the major constituents and the trace elements.

American research workers[104] found that in the analysis of alloys with a Q-switched ruby laser the ratio $I/V$ decreases with increasing volume because the line intensity goes up at a slower rate than the calculated crater volume.

Also in an analysis of steel carried out with the Zeiss model LMA 1, using a neodymium glass resonator without Q-switch, the volume increased faster than the intensity.[131] Hence the possibility is given of achieving higher total intensities by vaporizing a number of small portions of the sample region under test and combining the effects in a summary spectrum (factor $n^{0.61}$).[131] Investigations carried out at Zeiss, Jena, showed that for the alloy type 40 NiCrMo 15 the intensities of the lines Fe 359·7 nm and Ni 357·1 nm increase nearly linearly with increasing crater depth (Fig. 54). These findings can be interpreted to indicate that the quantities of various substances vaporized by laser action are projected into the gap between the two exciting electrodes within a cone of relatively small solid angle.

Through varying the pumping capacitance values and the inductance steps of the events sequence, the lifetime of the pumping source and of the laser resonator can be prolonged: reducing the capacitance of 508 μF, coupled with an inductance of 127 μH, to 381 μF, also with an inductance of 127 μH, and then further to 254 μF coupled with an inductance of 63 μH, appropriately changes the action rate of the events sequence and also the discharge cycle in the pumping circuit and the pulse characteristic of the coherent radiation.

The ratio of crater diameter to crater depth decreases with decreasing capacitance (Fig. 55). A capacitance of 254 μF gives, for special steels, a ratio of approximately 1 : 1.

Here too the intensity of the spectrum lines is not proportional to the volume of the vaporized sample material. Pumping with low capacitance

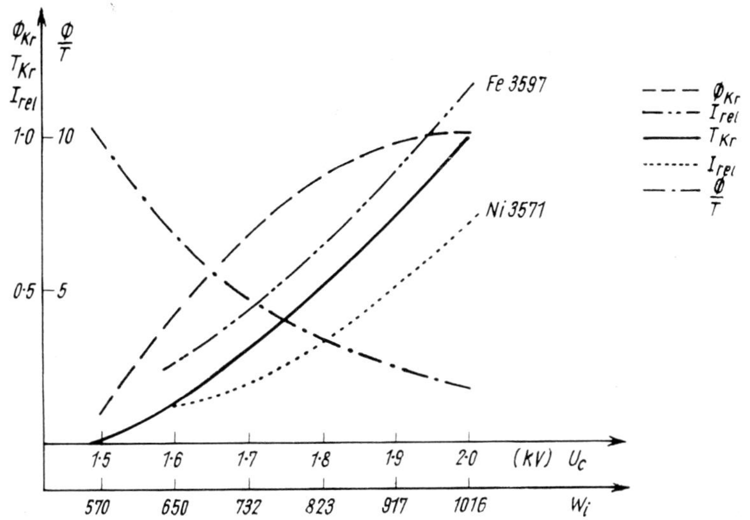


FIG. 54. Change of crater diameter $\phi_{Kr}$, crater depth $T_{Kr}$, ratio crater diameter/crater depth ($\phi/T$) and of spectrum line intensity $I$ with change of pumping voltage $U_c$ and energy $W_i$.

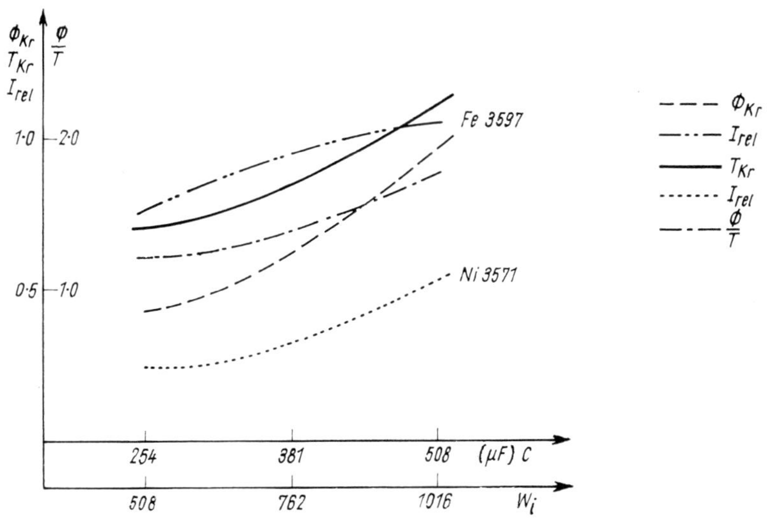


FIG. 55. Change of crater diameter $\phi_{Kr}$, crater depth $T_{Kr}$, ratio crater diameter/crater depth ($\phi/T$) and of spectrum line intensity $I$ with change of pumping circuit capacitance $C$ and pump energy $W_i$.

at high voltage is beneficial for both the xenon flash tube and the resonator; it is better to vaporize the sample with 2 kV/254 μF than with 1·7 kV/508 μF. The laser energy should only be reduced with one of the different diaphragms of an optical adaptor; for decreasing the capacitance, as just explained, does not have the desired effect (Fig. 56).

For craters in steel, the dimensions of which were only varied through reducing the divergence, the ratio of diameter, $d$, to depth, $h$, in the case of the largest volume has been reported as 2 : 1; in very small craters it becomes about 1 : 1.[131] With the Zeiss model LMA 1 the variations of these dimensional values diminish with smaller diaphragms.

### 3.3.1. Choice of operating modes with the Zeiss model LMA 1

The various alternative instrumental components and the working parameters available with the Zeiss model LMA 1 are shown diagrammatically in Fig. 57. The basic outfit includes a neodymium glass laser without Q-switch; the latter can be added as an accessory. Another optional addition are ruby resonators rods with or without Q-switch (not included in the basic standard outfit). The first selection to be made in preparation for the analysis is that of the objective: smaller sample quantities are always vaporized with the reflecting objective. Without using the optical adaptor between the microscope and laser head, the second possibility of obtaining a smaller crater diameter is reducing the battery capacitance in three steps. The third method of restricting the extent of the irradiation effect is offered by the continuous control of the flash tube voltage (Table 11).

TABLE 11. Choice of flash tube voltage for obtaining a crater radius corresponding to sample size

Material: Wolframite (MnFe) $WO_4$
Outfit: Model LMA 1, laser without Q-switch, glass resonator, without optical adaptor, and fixed battery capacitance of 508 μF

| *Flash tube voltage* (kV) | *Crater radius* (μm; ± 5 per cent) *Reflecting objective* × *40/0·50* | *Flat-field achromat* × *16/0·20* |
|---|---|---|
| 2·0 | 90 | 125 |
| 1·9 | 80 | 115 |
| 1·8 | 70 | 105 |
| 1·7 | 60 | 95 |
| 1·6 | 50 | 85 |
| 1·5 | 40 | 75 |
| 1·4 | 30 | 65 |
| 1·3 | 20 | 55 |

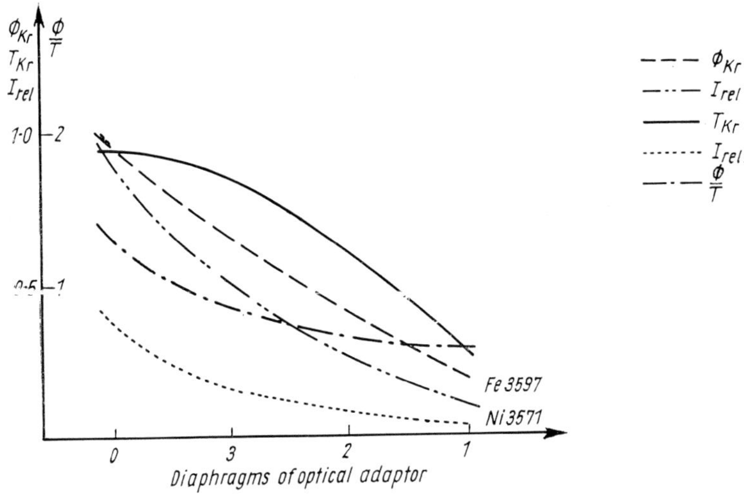


FIG. 56. Effect of diameter of divergence diaphragm on crater diameter $\phi_{Kr}$, crater depth $T_{Kr}$ and their ratio $\phi/T$ and on spectrum line intensity *I*.

0: no diaphragm in beam – 3, 2, 1: diaphragms of optical adapter in order of decreasing diameter.

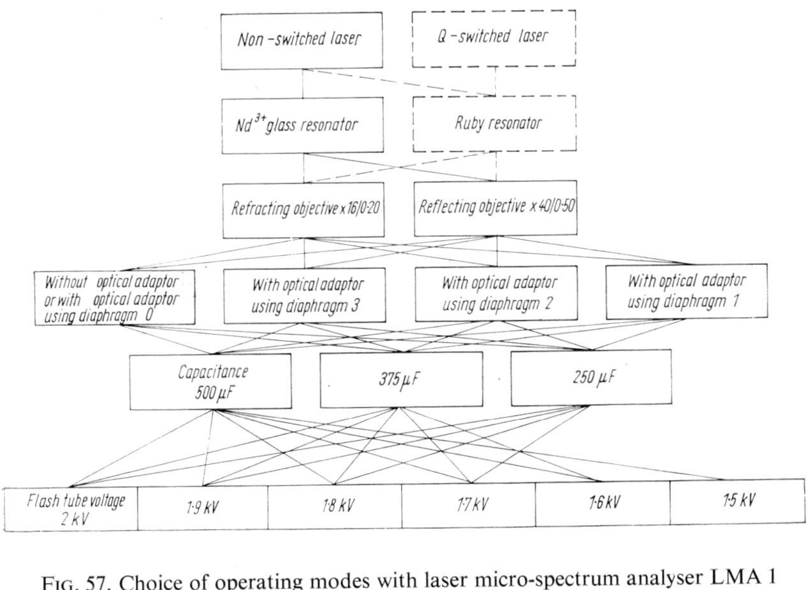


FIG. 57. Choice of operating modes with laser micro-spectrum analyser LMA 1 (Q-switch attachment and ruby resonator are not included in the standard outfit).

The optical adaptor provides a further, fourth, set of alternative operating conditions. One of three diaphragms can be brought into the light beam in the focal plane of the adaptor; their diameters are in the ratio of 4 : 2 : 1. With the smallest, used with appropriate settings of flash tube voltage and battery capacitance, crater diameters of 10 μm can be realized.

TABLE 12. Working modes for analysis of a 70 μm ore inclusion (1 to 3 with the standard outfit model LMA 1)

| | *Laser* | *Resonator* | *Objective* | *Battery capacitance* (μF) | *Flash tube voltage* (kV) |
|---|---|---|---|---|---|
| 1 | Without Q-switch | $Nd^{3+}$ glass | ×40/0·50 | 508 | 1·55 |
| 2 | Without Q-switch | $Nd^{3+}$ glass | ×16/0·20 | 508 | 1·3 |
| 3 | Without Q-switch | $Nd^{3+}$ glass | ×16/0·20 | 254 | 1·65 |
| 4 | Without Q-switch | Ruby | ×40/0·50 | 508 | 1·6 |
| 5 | Without Q-switch | Ruby | ×40/0·50 | 381 | 1·9 |
| 6 | Without Q-switch | Ruby | ×16/0·20 | 508 | 1·55 |
| 7 | With Q-switch | Ruby | ×40/0·50 | 508 | 2·0 |

The laser micro-spectrum analyser of Carl Zeiss, Jena, in its standard form gives the choice of about 190 different working modes; these will yield in different ways, according to the nature of the sample material, a certain crater diameter between 10 and 250 μm (Table 12).

The maximum capacitance of 508 μF is chosen for the analysis of substances that are difficult to vaporize, and for single excitation of a rather large sample mass (with objective ×16/0·2).

## 3.4. SAMPLE PREPARATION, INSTRUMENT SET-UP, PROCEDURE OF MICRO-ANALYSIS AND INTERPRETATION OF SPECTRA

### 3.4.1. Sample preparation

Different from the requirements for conventional emission spectroscopy with photographic spectrographs or direct reading instruments, the sample preparation for analysis with the laser probe is very simple. In many cases no preparation at all is required. Considerable time is therefore gained and the notorious errors due to contamination with foreign elements are avoided; these elements are often introduced in sample preparation through the use of reagents, mortars, ball units and auxiliary electrodes, through chemical reactions taking place in the bore holes of auxiliary electrodes and through mixing the sample with powdered carbon and matrix materials.

In most cases the sample is suitably secured on the microscope stage with clips or retaining strips. Simple objects like small pieces of minerals, crystals or alloys, wires or sheet metal, can be fixed on a supporting glass plate with plasticine.

If inclusions in transparent host materials are to be analysed, for instance chlorite, rutile, hematite, tourmaline or similar minerals in a quartz matrix (rock crystal), a flat surface is ground and polished on the sample in the appropriate direction until a fraction of the inclusion is exposed; the laser radiation is then focused on this fraction. Such ground and polished sections of ores, minerals, metals and alloys are prepared in the usual manner following well-established practices of mineralogical and metallographical institutes. Thin slices are similarly treated and should, when finished, be only partially protected by a cover glass. For larger samples, for instance antique ceramics and metal objects from archaeological sites etc., the microscope stage and possibly also its base have to be removed; these samples have to be clamped on a stand, independent of the laser microscope, best behind the optical bench of the spectrograph.

Solutions and powders, according to Rosan,[106] are dispersed in the gelatine layer of an undeveloped spectral film (Kodak 649-0 35 mm). Recently compressed disks (pellets) also have been used with success (see § 4.3). Single grains of a silver alloy of 40 to 50 μm diameter have been held in position with an adhesive tape.[103]

## 3.4.2. Instrument set-up

### *3.4.2.1. Setting up the LM microscope—Preparing the complete LMA 1 assembly—Adjusting and centring the microscope and the laser head*

After removal of all transit fixtures, and after the clamping screws of the interchangeable adaptor base (lower section of the LM unit) have been sufficiently withdrawn for the microscope to pass freely over the bench profile, the microscope itself is rested on the optical bench of the spectrograph. It is positioned at the correct distance from the spectrograph entrance slit and then secured by means of the clamping screw of the base.

If flat samples are to be examined in reflected light and if transparent samples have to be analysed with the use of transmitted light, then the condenser substage is inserted at the lower end of the microscope stand dovetail and the stage mount at the upper end, and both parts are secured in position with a key. The wide-angle lens condenser 0·6 for transmitted light is pushed home into the sleeve of the substage and fastened with the clamping screw. The position of the stage mount along the dovetail rail can be chosen so as to allow the accommodation of larger samples to be investigated by reflected light.

The optical adaptor is only added when focal spots of very small diameters are to be produced.

Once the mains supply unit has been set up, the microscope LM can be aligned.

### *3.4.2.2. Safety measures*

After publication of numerous pseudo-scientific descriptions in which laser radiation with its space and time coherence has been explained in fantastic terms such as 'death rays', a trend has more and more established itself to keep to facts on a scientific basis. The problem of precautions that may be necessary in the use of this new source of radiation has now to be considered.

The analyst must first of all be protected during three operations:

1 Safety must be assured for manipulations inside the laser head, for example changing the resonator or checking the condition of the pump flash tubes. This protection is provided by electrical safety switches: on opening the resonator housing the capacitor banks of the console are discharged.

2 During the viewing of sample details in the microscope no laser radiation must be reflected by the sample surface or enter the eyepiece by any other way.

In the Zeiss LMA outfit a unit is incorporated between the resonator and the eyepiece tube which is operated by a three-position selector lever on the side wall. Only with this selector in its centre position can the laser radiation from the resonator enter the microscope (type Nf); but in this setting the deviating prism is not in the light beam and therefore the sample cannot be viewed while it is irradiated by the laser radiation and no harm can be done to the operator's eye. When the microscope is set for visual observation a metal shutter is interposed between resonator and deviating prism. If the laser is inadvertently started the parallel light beam will only slightly damage the paint on the shutter.

The American microscope laser model 190 of Optics Technology Inc. has several shutters incorporated in the laser head; these prevent inadvertent discharge of the capacitor banks, and so also laser irradiation during sample adjustment and observation, and, furthermore, reflection of laser radiation into the eyepiece.

It has also been claimed that a red or infra-red filter in the beam from a small solid-state laser, mainly used in irradiation procedures, especially in biological and medical applications, allows visual observation of the laser impact without any risk.

Experience with highly reflecting samples, such as metals, ores, minerals, etc. has so far not been described in the literature. Safety is fully assured through the television technique in those rare cases of laser micro-spectroscopy in which the reaction of a substance to the laser radiation at the moment of the impact at the focal spot is to be investigated.

3 The spark gap is a possible danger source in laboratory work only if, with the electrode mains supply unit switched on, the microscope stage or the sample are handled without due care.

The Zeiss LMA model shown at the Leipzig Fair in 1965 already had insulated electrode holders and a safety switch on the mains supply unit II. There are several safety switches in the model LMA 1; on opening the screened microscope housing the capacitors of the electrode supply unit are automatically discharged.

The safety measures discussed in 1 to 3 are provided in all production models of the laser micro-spectrum analyser Mark LMA 1. A safety circuit is provided for the complete outfit that causes the laser capacitor bank to be discharged by any one of the following actions: opening the resonator head; opening the door covering the sockets on the side of the movable console; opening the safety door over the four capacitance and inductance switches on the front of the capacitor bank chassis; operating the microscope control knob for resonator adjustment.

When the safety doors of the LM microscope are opened and when the setting of the lever on the side of the microscope housing (part 4, Fig.10) is changed from unrestricted passage of the laser radiation to visual observation or photomicrography, the capacitors for the auxiliary spark gap are discharged and the laser ignition is interrupted. Similarly, the 2 kV and 5 kV high voltages are switched off when one or several of the chassis are pulled out. When not completely enclosed, as for instance without the back wall of the electronic console or without the housing of the LM microscope unit in position, the LMA 1 outfit must be operated only when the standard norms for supervision by qualified personnel (in the German Democratic Republic TGL 200–06 B.1.7.8) are complied with.

A red warning sign on the left side of the resonator head indicates that there is danger from the intense radiation of the pump source and from reflections of the coherent radiation when the dust cover over the recess for the rapid change mount of the Q-switch attachment is removed.

The manufacturers of commercial laser microscopes have only little freedom of choice between providing safety of operation and convenience of use. In certain special cases devices incorporated for the safety of the user may somewhat incommode him. Here the responsibility rests on the scientist who must not allow a laboratory technician to carry out such spectroscopic work. If in such a case the apparatus is interfered with, the manufacturer has of course no responsibility. But it should be pointed out that problems of this nature do sometimes arise in practice.

### *3.4.2.3. Microscopy*

The microscope with a solid-state laser mounted on the optical bench of a spectrograph is equipped for the most frequently used observation techniques, if only for no other reason than that of best utilization of the spectrograph. The model LMA 1 of the VEB Carl Zeiss, Jena, has facilities for bright field illumination and for qualitative polarization methods, both for work with reflected and transmitted light.

These visual microscopy methods allow on the one hand adequate qualitative estimation of unprepared samples and on the other hand the repeated location, often very important, of analytically interesting small sample parts, that is the details that are the object of the analysis.

In general polished sample sections and polished thin slices should first be examined with a standard routine or research microscope before being mounted on the stage of a laser microscope. The scheme of operations therefore is as follows:

| A | Unprepared Sample | B | Polished Sample Section or Slice |
|---|---|---|---|
| 1 | Qualitative estimation with microscope of laser micro-spectrum analyser | 1 | Qualitative and quantitative examination with standard microscope and marking of sample |
| 2 | Photomicrography | 2 | Photomicrography before and after mounting of sample on laser microscope stage |
| 3 | Micro-spectrochemical analysis | 3 | Micro-spectrochemical analysis |
| 4 | Examination by microscope and photomicrography and photographic recording of effect of laser impact | 4 | Microscopic examination and photomicrographic recording of laser crater |

The qualitative microscopic estimation of procedure stage A 1 may be supplemented with a hand-drawn sketch of prominent sample details.

Isotropic and anisotropic optical substances can be distinguished in the direct light beam between crossed polarizers with the help of the rotating stage of the LM microscope. When an anisotropic crystal plate cut at an angle other than 90° to the optical axis is rotated through 360°, the field of view darkens four times, namely in the positions in which the vibration directions in the plate are the same as those in the polarizer or analyser. The cross lines of the eyepiece graticule have of course to be set parallel to the principal planes of the polarizer and analyser. The optical orientation (orientation of the indicatrix relative to the crystal) is found from the orientation of the extinction positions relative to the available reference directions (crystal edges, cleavage structures, twinning lines). The setting of the graticule cross lines is controlled with a cleaved reference standard of anhydrite. In the preparation for micro-spectrochemical analysis one is usually interested in carrying out length measurements in the $x$–$y$ plane of the sample. For this purpose eyepieces are required that take micrometer scale graticules and that have an adjustable eyelens. The eyepiece micrometer scale of the first prototype of the LMA microscope series had to be calibrated for each eyepiece-objective combination in terms of absolute length units. This can simply be done by using a micrometer with the sample stage. The calibration value, that is the absolute length value corresponding to one scale unit of the eyepiece micrometer, is the quotient of the fraction of the number of scale units of the sample stage micrometer ($m_u$) into the corresponding number of scale units of the eyepiece micrometer ($m_0$). If for example one division of the sample stage micrometer scale is 0·01 mm and if the dimension of a sample detail being measured covers, as seen under the microscope, $t$ divisions of the eyepiece micrometer scale, then the true absolute length of the sample detail is

$$l = (m_u t/m_0) \times 0{\cdot}01 \text{ mm}$$

Polarized light interference patterns cannot be satisfactorily obtained in the indirect (conoscopic) light beam with a reflecting optical system because of such a system's central vignetting. Therefore laser microscopes for the LMA model have no Bertrand lens. Standard polarizing microscopes have to be used for investigating the optical properties of samples in the indirect light beam. The optical characteristics of thin slices of trigonal, hexagonal and tetragonal crystal samples, cut parallel or obliquely to the optical axis, can be determined in the direct light beam by means of a waveplate for first order Red. This is done by inserting the crystal plate under test at 45° (diagonal position) between the crossed polarizer and analyser and noting whether the introduced waveplate likewise orientated at 45° changes the interference colour by a shift towards the higher orders or to the colour corresponding to the subtraction (Red $-$ $\lambda$).

The determination is especially simple when the direction of the principal symmetry axis is revealed by the morphology of a small crystal (for example elongated in the direction of the $c$-axis, tabular towards the base). The optical nature of a thin crystal slice is deduced from the relative values of the refractive indices. The ordinary ray in uniaxial crystals always vibrates at right angles to the principal crystallographic axis; if this direction corresponds to the higher velocity of the light propagation, that is to the smaller refractive index, then the crystal is optically positive; otherwise it is optically negative.

In the case of addition (Red + $\lambda$) the waveplate changes the first order Grey of certain minerals (very low birefringence or very thin slice) into the intense interference colours at the boundary between first and second order (deep red and purple red go over to the violet to blue hues; 'sensitive tint').

If required, pleochroism of sample components is examined with the polarizer in the light beam, and specified with reference to standard directions that can be clearly described and easily located; these should, if possible, be principal vibration directions of a crystal—or otherwise some morphological orientation like cleavage structures.

The images seen in the LMA 1 model are erect and not reversed horizontally, a feature considered as a concession to convenience of use.

Preliminary investigations can be carried out with magnifications ×32 and ×50. The refracting objective ×40/0·05, ∞/– * provided for this purpose and also the reflecting objective and the objective ×16/0·20 for semi-micro-analysis have a parfocalizing distance of 75 mm. Two objectives are mounted on a selector slide for rapid change which is a significant improvement compared with the laboratory model LMA.

It is important for the user to know that a plane-parallel glass plate that can be withdrawn serves for providing the bright field illumination for work with reflected light and that the reflecting objective ×40/0·5,∞/0 is working with true Köhler illumination.

## 3.4.3. Procedure of micro-analysis and interpretation of spectra

### *3.4.3.1. Obtaining the spectrum*

Different from the usual practice of emission spectroscopy with a spark or arc source, in laser micro-spectrochemical analysis the sample does not form the electrode and is also not placed in a cupped electrode. In the two-step method (cross-excitation process) (see § 3.1) the ionized material ejected from the sample sets off a capacitor discharge between two main electrodes, preferably of spectrum carbon. For reason of symmetry the electrodes have identical diameters and identically shaped tips; a length of about 50 mm, a diameter of 5 mm and a tip of 10° angle are recommended. The electrodes should preferably be set up so that their tips are inclined to the sample surface at a certain angle (70° ± 10° adjustable in the LMA model) and that their axes are at right angles to the axis of the illumination system and therefore—after alignment—to the collimator axis of the spectrograph.

After the sample has been appropriately mounted on the microscope stage its surface, ideally plane, lies parallel to the table top and at right angles to the spectrograph slit.

The generation of the spectrum begins with the short duration process of vaporization of the sample through the focused laser radiation and excitation of the ionized plume in the spark gap between the sample and the front plate of the objective. Before this process is started the spectroscopic light source, the adjustable illumination system and the centre of the spectro-

* See footnote p. 48.

graph entrance slit have to be aligned on a common optical axis parallel to the axis of the optical bench of the spectrograph. Then this axis is made to coincide with the collimator axis.

For the alignment procedure in laser micro-spectroscopy the instructions for conventional emission spectroscopy can be followed. Required are a system for electrode projection, a spirit level, in a mount of 80 mm maximum length if it is to be used on the spectrograph model Q 24, two condensers on stands for the optical bench (the single quartz lenses for the ultra-violet of 80 mm and 160 mm focal length respectively with a diaphragm wheel have been found especially useful), and a centring mount with circle and cross lines graticule. The instructions established in conventional spectroscopy for the spark stand here do not apply unchanged: The setting screws mentioned in this connection in the instructions are in the LMA model found on the electrode holder of the microscope; with the distance from the electrodes to the centring ring on the spectrograph slit set to 50 cm and that to the imaging condenser ($f$ = 80 mm) to 10 cm, the enlarged image of the discharge can be aligned with these settings screws symmetrical about the circle of the centring mount both in the horizontal and in the vertical direction. If on moving the quartz condenser to within about 10 cm from the spectrograph slit the resulting reduced image of the spark discharge does not fall exactly on the point of intersection of the centring cross lines, the necessary correction is made in the usual way by means of the knurled head screws of the condenser.

The electrode tips, viewed with the large-field objective $\times 4$ and with the diaphragm of the condenser for reflected light closed down, can with the two setting screws on the holder be adjusted symmetrical about the cross lines of the microscope eyepiece; the range of this adjustment on the LMA 1 model is $\pm 2$ mm vertical and horizontal in the direction of the collimator axis. This is always done at the start of the alignment procedure. Similar to all arc and spark stands the Zeiss laser micro-spectrum analyser can be rotated in the $x$–$y$ plane; the microscope with the adjusted resonator housing mounted on it can be turned through small angles on its base, which is clamped rigidly on the optical bench.

The optical axes of the illumination system and of the spectrograph are brought into coincidence in the usual way with the use of the three condensers for intermediate imaging, the slit being wide open and the camera objective observed visually.

After the alignment the two condensers of 160 mm and 200 mm focal length, not required for qualitative analysis, are removed and the microscope and the condenser of 80 mm focal length are brought into their proper positions on the optical bench and the discharge is imaged on the slit.

A shutter placed between condenser and spectrograph slit is very useful; it makes it unnecessary to operate repeatedly the cassette slide when the material for producing the spectrum is vaporized by several laser pulses. For analysis of samples of unknown composition the supply unit for the spark gap is set to medium values, that is for instance capacitance 12·5 μF, inductance 125 μH, voltage for carbon electrodes 2·5 kV. If the CN band at 388 nm has to be suppressed as effectively as possible the inductance is set to the lowest value available on the supply unit. With high inductance, of about 1000 μH or more, the capacitor discharge runs almost normal to the spectrograph slit; to compensate for that effect one will always work with maximum capacitance and observe if the radiation beam enters the optical system of the spectrograph under the best possible conditions.

The electrode tips are usually contaminated through inadverted touching with the fingers or by the packing material; it is therefore advisable to clean them before mounting the sample on the microscope stage by discharging the capacitor several times at a capacitance of 2·5 μF and at maximum supply unit voltage.

Errors in the results of the analysis can stem from any of the following causes:

Sparking between the electrodes and the sample surface. This can be avoided by increasing the separation.

Contamination of the quartz protecting plate of the reflecting objective.

Vertical displacement of the electrode tips with respect to the slit centre, causing tapered spectrum lines.

Too great a distance between the two electrodes or between electrodes and sample surface.

Too low an inductance for pure arc lines and too high an inductance for spark lines.

Too low an output energy of the laser.

Partial obscuring of the spectrograph slit by parts of the sample.

After the cross excitation a thin oxide layer is formed on certain samples as a result of the capacitor discharge in the spark gap, especially on metals and polished sections of ores and alloys. This affects the subsequent microscope examination of the structure and crater. In such cases a special masking technique is used in which a piece of carton with several holes of about 0·5 to 1 mm diameter is placed on the sample. The laser beam is focused through one of the holes on a detail of the sample surface to be analysed. The synchronized excitation process then takes its course without disturbance. The authors adopted this excitation method in the examination of platinum ores and of silicate inclusions in steels.[103] It is advisable to analyse a blank sample first so that chemical elements possibly present in the mask material can be taken into account in the interpretation of the results.

Cross excitation of materials vaporized by laser irradiation close to biological samples causes complications because the masking technique usually fails in investigations of such substances, as for instance blood corpuscles. A region around the laser crater itself is affected by the irradiation which analytically cannot really be evaluated. According to the literature[11] damage to the periphery of biological and medical samples can be almost completely avoided by using electrodes of a special form mounted in an adjustable holder and by keeping the parameters of operation of the cross-excitation stage low, that is at about 2 μF, 0·8–1·2 kV, 20 μH, 6·6 Ω; in this method spark energies of 1 J are discharged in a critically damped form between bevelled graphite electrodes (pointed to 1 mm and then chamfered), with a separation of the plane-parallel electrode segments of 0·2 mm, an inclination to the sample surface of 30° and a distance of 1 mm between electrode tips and sample. The laser output, attenuated by a $CuSO_4$ solution, was 0·5 to 10 mJ.

Theoretically graphite rods can be employed instead of carbon electrodes as this example proves.

In the laser microprobe of Jarrel-Ash, thin electrodes of 0·125 inch (3·2 mm) diameter are used; an excitation voltage from 1·5 kV upwards can be chosen; the electrode separation is 2 mm. Maxwell[75] in his investigations of mineralogical samples used a single spherical condenser for imaging the spark discharge (without additional inductance). Rosan[106] preferred excitation with 1·6 kV, 20 μF and 50 μH, whereas Swaine[125] gives as suitable values 1·5 kV, 20 μF and 100 μH.

In work with the spectrograph model Q 24 the authors basically only use the aperture ratios f/10 or f/15; with the latter value the spectrum lines are sharp and well defined throughout the whole range from 200 to 650 nm.

When choosing photographic spectrum plates one has to remember that the general sensitivity increases for the different types in the order from 'Ultrahard' to 'Extrahard', 'Hard' and 'Rapid', but that the resolution decreases in the same way.

For routine analyses with the LMA model the authors employed ORWO plates WU3 because of their fine grain. These plates are sensitive up to a wavelength limit of about 500 nm whereas the working range of orthochromatic WO spectrum plates extends to 580 nm. Since 1967 the authors have also obtained very good results with the extremely sensitive ORWO Astro Special plates ZU 2. They are developed for 5 minutes at +20 °C in ORWO 70 or in A 71 developer.

Rosan[106] treats 103-0 Kodak film with D-19 developer and also recommends 'Royal-x-Pan' film. Laqua (private communication) uses English Kodak 10 and Ilford Selochrome plates.

Maxwell[74,75] found in experiments with the microprobe of Schuch and Brech (Jarrell-Ash) the 'Kodak 1030' 35 mm film preferable to other types; Kodak film '1 N' was unsuitable.

D.J. Swaine[125] used with the Hilger quartz Littrow spectrograph Ilford XK plates for the range 248–360 nm and HP 3 emulsion from

350 to 650 nm. His insistence on the necessity of air transport of the highly sensitive spectral plates and on their storage under refrigeration holds equally for other similar materials, such as ORWO ZU 2.

Karjakin *et al.*[58] used GOST standard plates Type II 16 with the spectrograph model ISP-22.

Korolev *et al.*[64] employed Panchrom 10 A film with a sensitivity of 1500 units of the GOST standard 0·85. In Japan the team of Hitachi and the Municipal Research Institute of Tokyo achieved good results with Fuji film emulsion of ASA sensitivity 700.

For photographs to be taken for preliminary inspection with the LMA outfit and the Zeiss standard glass spectrograph, the plates ORWO-Red-WP-1 and ORWO-Red-WP-3 (rapid and extra hard) are available; and in addition the infra-red plates I 850 with a peak sensitivity at 850 nm.

When used with the aperture ratio f/30 the spectrograph model Q 24 requires twice as many laser pulses as for the aperture ratio f/11 in order to give comparable spectrum lines of approximately the same photographic density. Similarly more laser pulses per spectrum (three to five times as many) are necessary for quantitative than for qualitative analyses; in the former, intermediate images have to be formed by complete optical systems of components with focal lengths of 80, 160 and 200 mm or by the new quartz-lithium fluoride achromats for the ultra-violet of 75·8 mm and 150 mm focal lengths, together with the 200 mm field lens attached to the spectrograph slit mount.

### *3.4.3.2. Taking a spectrum photograph for preliminary inspection with the LMA model*

The conditions essential for obtaining the best analysis results are: adequate alignment of microscope sample stage, objective, condenser for transmitted light and optical adaptor, and exact adjustment of the laser resonator. Furthermore the electrodes must be properly set relative to the optical axes of the microscope and of the spectrograph.

For qualitative work with the Q 24 quartz spectrograph the following operating conditions are recommended: imaging the electrodes with the quartz condenser of 80 mm focal length directly on the slit of 12 to 20 μm width and 2 mm length; camera aperture ratio *f*/11 or *f*/15; optimum collimation.

Corresponding suggestions for the PGS 2 plane grating spectrograph are: direct electrode imaging as with the Q 24 spectrograph, slit width 20 μm, slit length 2 mm (step setting 3 of the automatic cassette drive), grating with 325 lines/mm and blaze wavelength of 300 nm, grating mask 70 × 50 mm, single pass. Grating angle, collimator and slit tilt are set according to the instructions for the selected grating.

Laser capacitor bank 387 μF; flash tube voltage 1·9 kV. Microscope spark gap: capacitance 2 μF, inductance 60 μH, voltage 2·2 kV. Electrode separation of about 0·5 mm, that is with an objective ×4/0·05 and an eyepiece ×12·5, about twelve scale divisions on each side of the centre point of the eyepiece graticule.

*Final check before start of operation*

1 The charging voltage $U_L$ for the ZFL 1000 laser flash tube is first set to 0 kV. To find the breakdown voltage for the spark gap $U_E$ the mains supply is switched on and the control knob is turned to the right until a discharge occurs between the electrode tips; this also cleans up any contamination. The reading of the corresponding voltage value $U_E$ in kV on the meter scale is noted. After this discharge the voltage $U_E$ is set to a value just below the breakdown voltage so found. The mains supply of this chassis (2) for the auxiliary electrodes is then switched off.

2 The microscope illumination is switched on. The sample surface viewed under the microscope is set up at a distance of about 0·8 mm from the electrode tips.

3 The microscope is ready for the analysis when the lever (4) points backwards for unimpeded passage of the laser radiation, the safety doors are closed and the index line of the control knob (14) points upwards or to the right. On setting the lever (4) the microscope illumination is automatically switched off so that the spectrum plate cannot be inadvertently exposed.

4 The electronic console is ready for use when the following indicator lamps are lit: the right indicator lamp (for the laser supply) of the laser operation chassis (part 1 in Fig. 5); the two indicator lamps for the high-tension voltage and discharge start of the laser flash tube charging chassis (3); the two indicator lamps for the microscope illumination and for the spark gap supply of the chassis serving the cross-excitation electrodes (2). About 60 seconds after switching on, the capacitor voltage $U_L$ can be set and the laser can be fixed, either by the release on the console or by the hand switch according to the setting of the selector switch. The capacitors are automatically recharged after the two-step process of vaporization and cross excitation. When a single analysis is to be carried out it is therefore advisable to set the flash tube voltage control up against the left stop before starting.

Inherent in the design of the LMA 1 model is a starting delay of about 0·7 s. For repeated consecutive pumpings the cooling must be switched on when altogether six indicator lamps of the console will be lit.

5 Routine manipulations in the analysis procedure:

(a) The cassette and the spectrograph shutter are opened.

(b) After selection of the capacitance and inductance values the flash tube voltage $U_L$ is set to a suitable value which lies between 1·5 and 2·0 kV.

(c) The electrode voltage $U_E$ is allowed to rise to a value about 30 V below the breakdown value found experimentally before.

(d) The laser is fired.

(e) Both voltage switches are moved back to the left-hand stop.

(f) The spectrograph shutter is closed.

(g) The lever (4) is reset.

(h) The effect of the laser irradiation on the sample is examined visually through the binocular eyepiece.

(i) The crater diameter is measured with the eyepiece graticule scale. When the reflecting objective is used the real linear dimensions in micrometres are obtained by multiplying the scale reading (number of divisions) by two. With the flat-field achromat ×16/0·2 the factor is five.

(j) The focusing fine control is used for measuring the crater depth. One whole revolution corresponds to 100 μm and one scale division to 2 μm.

(k) The microscope doors are opened; the electrodes are changed, the replacement electrodes aligned and their tips decontaminated by several capacitor discharges; the sample is re-orientated for focusing on another detail.

(l) The spectrograph cassette is transported unless several laser firings are necessary for an analysis of minor constituents.

(m) The lever (4) is moved back.

(n) A new cycle of operations is started with operation (a)

### *3.4.3.3. Interpretation of the spectrum*

The cross-excitation process involving the two steps of sample vaporization and plasma excitation was soon adopted in practical work because, amongst other reasons, in spite of the use of carbon electrodes no interference by cyanogen bands can be detected. In the spectrochemical analysis of dielectric materials (ceramics, silicate minerals and others) with spectrum carbon electrodes, as carried out so far, numerous cyanogen bands are included in the spectrum region from 350 and 460 nm. Therefore medium quartz prism and small glass prism spectrographs cannot as a rule be utilized to full advantage over their entire wavelength range.

Different from continuous or intermittent arc spectra taken with exposure times from 10 seconds to several minutes, spectra recorded on plates such as type ORWO WU 3 with a laser micro-analyser in less than a millisecond are exceptionally sharp. In every spectrum (see for instance Fig. 58 and 59) various carbon lines can be seen, for instance C I 247·8 nm, C II 426·7 nm, 250·9 nm, 251·2 nm and C III 229·7 nm; but the whole gamut of interfering cyanogen molecular bands does not show

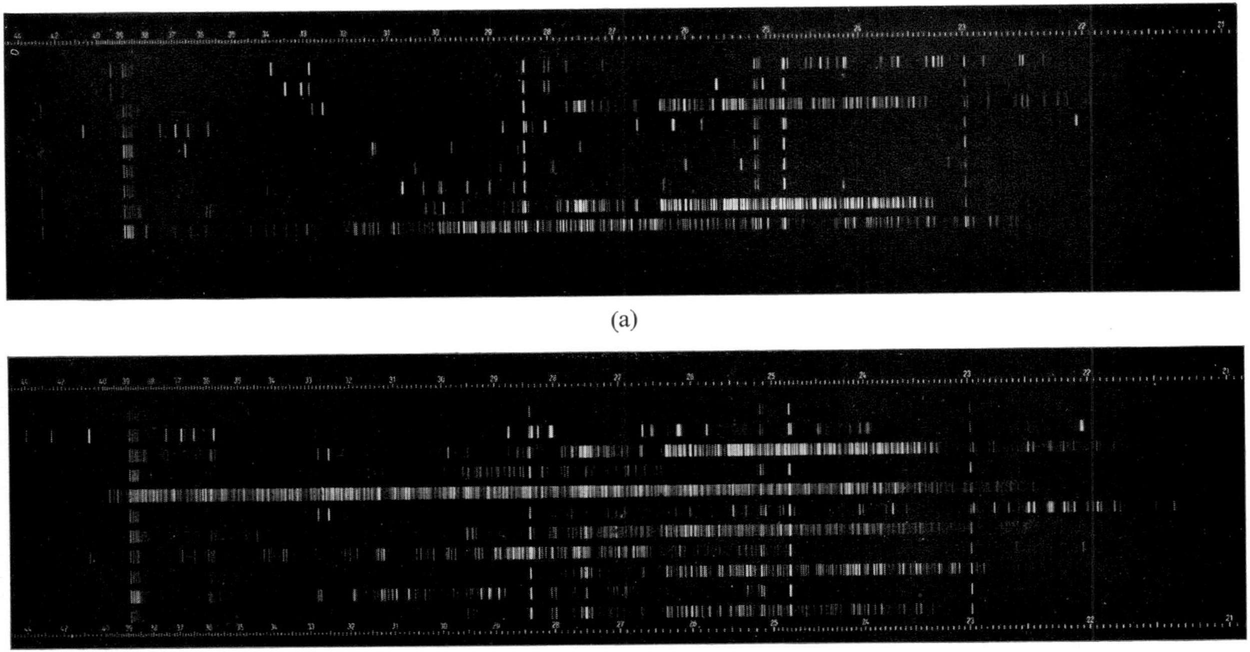

(a)

(b)

FIG. 58. Spectra taken with the medium quartz prism spectrograph model Q 24.

(*a*) Micro-analysis of sulphide ores with the LMA outfit. (Vaporization by solid-state laser model ZFL 750 fitted with ruby resonator; output energy 0·03 J).
From top to bottom: Argenite, sphalerite, chalcopyrite, galena, stibnite, bismuthinite, pyrite, molybdenite.
(*b*) Micro-analysis of various materials with the LMA outfit.
(Vaporization by laser radiation from $Nd^{3+}$ glass resonator, output energy 0·4 J, and excitation by auxiliary spark discharge.)
From top to bottom: Spectral carbon, galena (sulphide), chalcopyrite (sulphide), molybdenite (sulphide), samarskite (oxide), azurite (carbonate), wolframite (tungstate), crocoite (chromate), erythrite (arsenate), pucherite (vanadate), triphylite (phosphate).

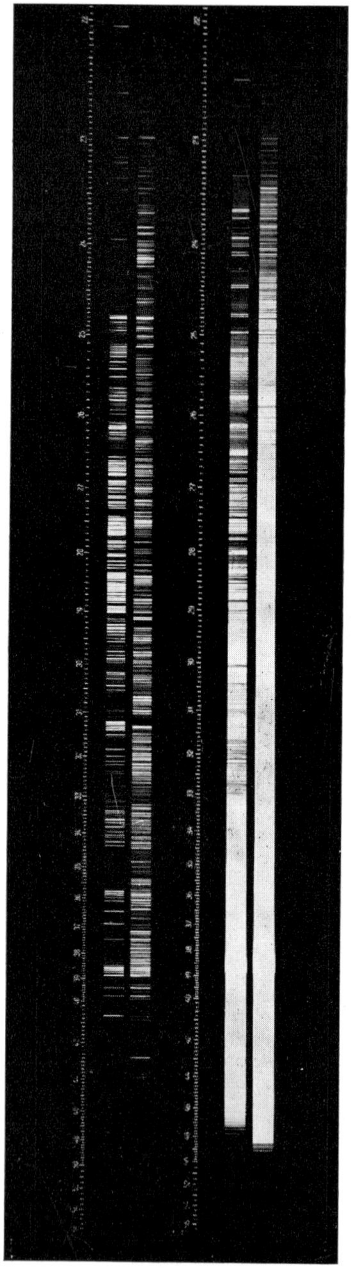

FIG. 59. Comparison of quartz spectrograph (Q 24) spectra obtained in laser micro-analysis with the LMA outfit and in conventional macro-analysis with a d.c. continuous arc.

Samples: Lead chromate (spectra 1 and 3) and samarskite (spectra 2 and 4). With the conventional type of excitation (spectra 3 and 4) considerable portions of the visible spectrum cannot be evaluated because of very intense band emission.

up. If the electrode tips are correctly aligned only one weak CN band between 380 and 389 nm appears (Fig. 60). This does not in practice complicate the determination of any element and spectrochemical micro-analysis of conducting and dielectric materials can be carried out throughout the wavelength range of a quartz prism spectrograph. The intensity of the weak CN band between 380 and 389 nm increases somewhat with higher inductance. In pure spark spectra taken with a capacitance between 1 and 3 μF it is hardly noticeable if only one laser pulse is used for each analysis. With a higher capacitance of 10 to 25 μF and with very sensitive spectrum plates, for example ORWO ZU 2 or Kodak I-0, the CN bands become more pronounced.

The qualitative assessment of the spectra can be satisfactorily undertaken on a spectrum projector with a magnification of ×20, which should also allow the examination of wet spectrum plates immediatey after the final rinsing; the model SP 2 of the VEB Carl Zeiss, Jena, for instance offers this facility and it also has a device for marking line pairs afterwards to be measured on a photometer. A great many principal spectrum lines used for identifying chemical elements can be found with the help of a standard iron spectrum in which analytical lines of other basic substances have been entered. Voigtsberger and Gössler compiled an atlas for medium quartz spectrographs, especially the Zeiss model Q 24 which has been published by the VEB Carl Zeiss, Jena, in a new enlarged edition in 1959.[129] The new edition includes for spectrographs of this type for the first time the rare earths Dy, Gd, Nd, Pr, Y and Yb; they all

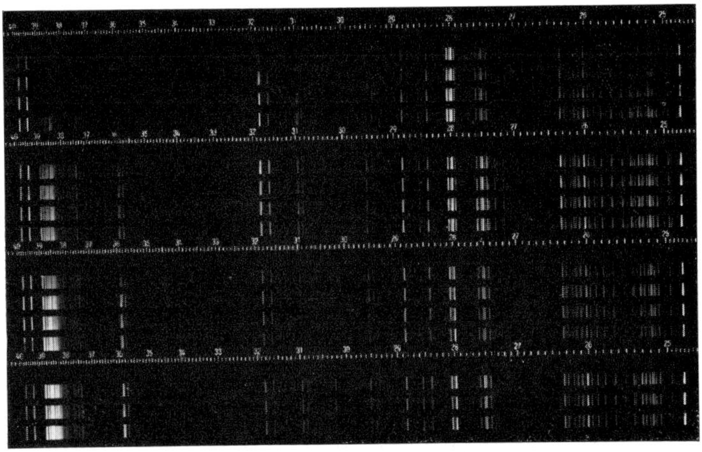

FIG. 60. Effect of increase in inductance on emission intensity of CN bands between 380 and 389 nm.

Sample: orthite.
From top to bottom: Increase of inductance from 0·025 to 1·12 mH.

can be determined in suitable samples with a laser micro-analyser. The authors, therefore, recommend the use of up-to-date spectrum atlases for routine work with a laser probe.

Understandably not all chemical elements could be included in the spectrum atlases and tables originally produced for conventional emission spectroscopy with arc or spark sources; a reference work with spectrum tables therefore is a necessary aid for the interpretation of spectra. On preference the authors used the following in their day-to-day work:

A.N. SAIDEL, W.K. PROKOFJEW, S.M. RAISKI, *Spectraltabellen*, (2nd edition, Berlin 1961)
L.H. AHRENS, S.R. TAYLOR, *Spectrochemical Analysis*, (2nd edition, London 1961)

A valuable addition to the Q 24 atlas of the iron spectra is the instructional spectra collection for glass spectrographs of S.K. Kalinin, L. E. Neumark, W.L. Marsuwanow and K.I. Ismagulowa, published in 1956 by the Kazakhstan Academy of Sciences at Alma Ata. This atlas can be applied directly to the interpretation of spectra obtained with an outfit consisting of the laser micro-spectrum analyser LMA 1 and a standard glass spectrograph with collimator and camera of 70 mm focal lengths.

The photometric evaluation of line pairs in quantitative micro-analysis is carried out in the usual way with the photoelectric microphotometer, now supplied with a new transformation scale. This '*P*-scale' is based on the relationship

$$P = \tfrac{1}{2}(S + W) = \tfrac{1}{2}\log\left(\frac{1}{D^2} - \frac{1}{D}\right)$$

Its purpose is to extend automatically the characteristic curve for photographs taken in the ultra-violet range during measurement of spectrum line pairs with the microphotometer. For density $S = 0{\cdot}209$ and transmission $D = 0{\cdot}618$ the $P$ value is zero.

For quantitative work in the visible spectrum region with a glass prism or grating spectrograph the Seidel transformation is recommended

$$W = \log\left(\frac{1}{D} - 1\right)$$

It extends the lower part of the characteristic curve between the $S$ (density) values of about 0·5 and 0·04, resulting in a linear relationship between the plate density and the logarithm of the element concentration. All Zeiss microphotometers manufactured before 1963 have the $W$ scale of the Seidel transformation. The type of the scale can easily be ascertained by a simple test because the zero point of the $W$ scale corresponds to the value 30 of the logarithmic density scale '$S$' and to the value 500 of the linear transmission scale '$D$'.

Because a Q-switched laser not only vaporizes, but also excites the sample so that auxiliary spark excitation is not absolutely necessary, the appearance of the spectra is somewhat different. The spectra of the plasma vapour produced by a Q-switched laser show a strong continuous radiation extending upwards to about 300 nm which is not found in the spectra excited by the electric spark near the sample surface. The spectrum lines obtained from the centre of the plasma cloud are very much broadened and there is sometimes also self-reversal. These effects are caused by the heterogeneity of the plasma vapour cloud (which consists of a mixture of incandescent solid and fused particles) of excited and ionized atoms. An exact electron temperature defined in space and time is hard to state, a condition different from those in the electric spark. In the two-step process with cross excitation, on the other hand, the spectrum characteristics are completely determined by the secondary excitation: the lines are sharp, their intensity is ten times as high as that of lines produced without auxiliary excitation and they can be better interpreted.[88]

### 3.4.4. Upkeep and servicing

A laser micro-spectrum analyser should, if possible, only be operated in a dust-free atmosphere. The resonator must always be clean and has to be regularly checked. The resonator rod is never to be handled without clean cotton gloves and to be touched only on the cylindrical surface. This surface may be carefully wiped with a clean dry linen cloth or, if there are fingerprints or grease spots, methylated spirit may be used. Traces of organic substances on the laser rod or on the xenon flash tube should always be removed immediately because they decompose under the action of the intense ultra-violet radiation. Inside and outside of the protecting tube should be cleaned in the same way.

Dust particles that have settled on the exit surface of the resonator rod and on the roof prism are removed with a clean rubber blower. These surfaces must be examined with special care and any dirt adhering more firmly should be rinsed off with methylated spirit without rubbing the surfaces. If a coloured film remains at the edges after evaporation of the cleansing liquid then the resonator has to be recoated. The oblique, internally reflecting surfaces of the roof prism, which is very sensitive to mechanical shocks, must always be very clean. A coloured or grey film, visible particularly in reflected light, or accidental finger marks must immediately be removed with methylated spirit.

If the sample stage after prolonged use moves jerkily or too freely it has to be lubricated with the grease provided. To do this the threaded ring on the underside of the stage is released right up against the stop and the stage top is pulled off sideways. The slides are cleaned with methylated spirit and slightly and uniformly greased and the stage is then re-assembled.

After prolonged operation the air filter in the resonator head should be washed and dried. Dust settled in the laser head should be carefully removed with brush and rubber blower. The flash tube should be replaced in good time. The approach of the end of its life is indicated by deteriorating striking or by electrode evaporation which causes a deposit inside the tube. The front protecting plates of the objectives must be cleaned daily and the microscope stand should also be kept perfectly clean.

If the optical adaptor has been working for some considerable time and the diaphragms are found to be damaged by the laser radiation, they can

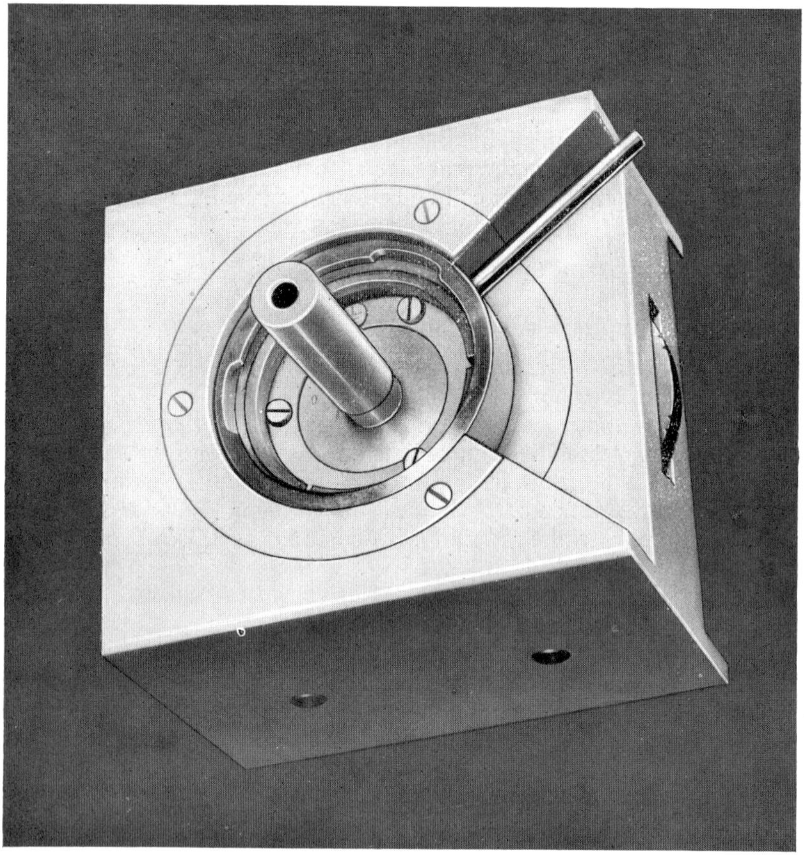

(a)

FIG. 61. Optical adaptor for the laser micro-spectrum analyser LMA 1 of VEB Carl Zeiss, Jena. (*a*) Total view. (*b*) Diagram.

1 Rapid change connector; 1a Three fixing screws for connector; 2 Telescope lens; 3 Diaphragm wheel; 3a Cap; 3b Spacing ring; 3c Diaphragm; 3d Adjusting screws; 4 Telescope lens.

be replaced by spares available on order. During this operation the optical adaptor is left in position on the microscope. The diaphragm wheel is adjusted. After removing the three screws (Fig. 61b, part (1a)) the whole assembly, part (1), is pulled out upwards and rested on the main body of the optical adaptor; the earthing cable need not be disconnected. The cap (3a) is unscrewed and the spacing ring (3b) and the damaged diaphragm (3c) are taken out. The new diaphragm is inserted and the unit is reassembled and secured by the screws. The control knob, (Fig. 10, part (14)) is set to the left and the new diaphragm is centred on the eyepiece graticule cross lines by means of the three adjusting screws (3d) (Fig. 61). The adjustment of the diaphragm wheel must not be disturbed in this operation. The assembly part (1) is then again fitted into the main body.

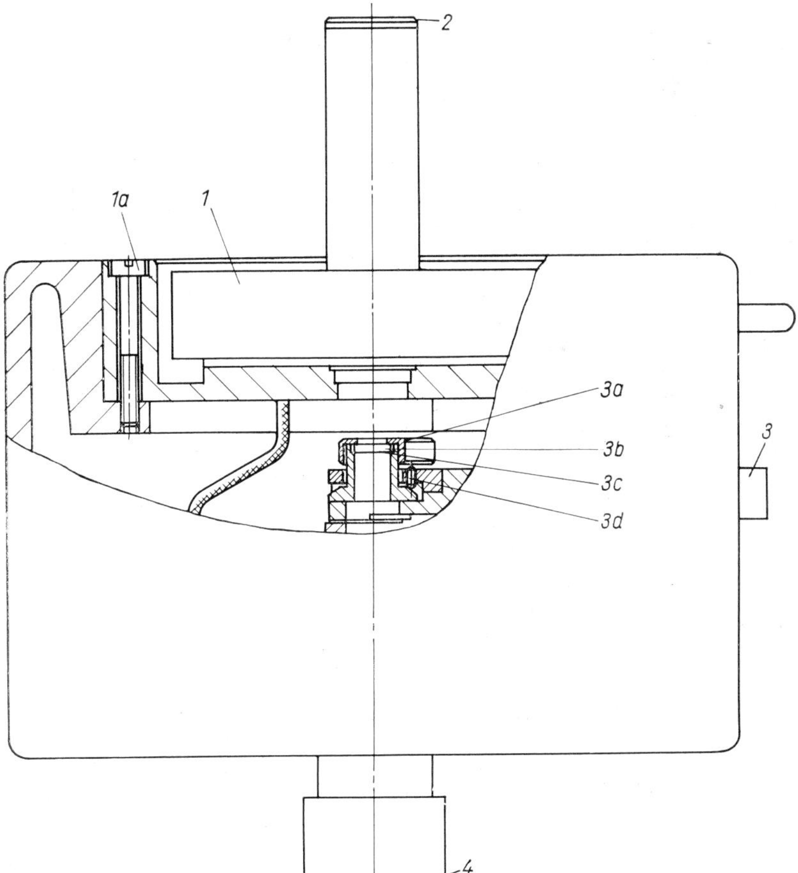


(b)

# 4

# Scope of the Method and Development Trends

## 4.1. COMPLEX FIELDS OF APPLICATION

This section gives information on complex fields of applications in various branches of natural science.

### 4.1.1. Mineralogy (prospecting in various types of deposits, petrography, investigation of meteorites) and geology

So far very careful and time consuming sample preparation has been required for the spectrochemical analysis of minerals and rocks.[79] In most cases the time necessary for this preparation was more than that available for the analysis itself, that is recording the spectra, processing the photographs and interpretation. The light source usually was a continuous d.c. arc; non-conducting powder samples were vaporized in hollowed auxiliary carbon electrodes. Vaporization and emission here occur in a single process. Conventional emission spectroscopy of mineralogical samples calls for high experimental skill and, because of the great variety of substances, for extensive experience and, unfortunately, still needs too much time.

Efforts have been made in the past to establish methods for spectrochemical spot analysis of polished sections and thin polished slices of samples; such methods have, however, always been applied only by a few analytical laboratories. In such a spot analysis the sample area under test is retained undisturbed in its immediate surroundings; the time required for preparation has already been shortened to a certain degree and the basis for genetic considerations laid down.

The simplest attempts at spectrochemical spot analysis were made without a microscope. Rubeska *et al.*[110] in Prague, for instance, in the investigations of conducting samples used as electrode a platinum wire of 0·3 mm diameter set in glass; and they used similar pointed electrodes, inclined at 45° to the sample surface, for the qualitative analysis of non-conducting materials. In qualitative analysis with the quartz prism spectrograph model ISP-22 from the USSR., a sample area of 0·5 mm diameter can be excited by a spark source at 8000 V and 500 pF, with an exposure time of 90 seconds.

For the quantitative analysis of iron, nickel and cobalt arsenides, with a mean square deviation of 14 to 18.5 per cent, the samples have to be treated locally with nitric acid (1 : 1); the resulting nitric solution is dropped on to carbon electrodes that have been immersed in sodium chloride; the electrodes are then excited by an a.c. arc discharge of 7 A for 60 seconds. Spectra of this quantitative spot analysis were recorded with a Hilger spectrograph model E 492.[110]

In the terminology of Loofbourow[68] this method is not one of micro-spectroscopy because it is was not carried out with the characteristic optical combination of microscope and spectrograph. Such a system for instance was chosen by Genkin and Koroljew in Leningrad[39] who assembled for the micro-spectrochemical analysis of ore constituents an outfit consisting of a polarizing microscope model MIN-8 with illumination for opaque samples model OI-12, a micro-manipulator MM-1, a sample guide ST-12 and a quartz prism spectrograph ISP-28 (Fig. 62). Various forms of micro-electrodes were developed for vaporizing the ore samples on the microscope stage and exciting them to emit spectral radiation. The condenser for work with reflected light has a 90° deviating prism which can be swung out of the light path; the radiation emitted by the sample also passes through the microscope objective, but in the opposite direction to the illuminating beam for visual observation. Achromats × 10/0·2 and × 5/0·08 for the ultra-violet were mostly used.

In 1967 Koroljewi et al.[72] developed at LOMO the model MSL-2 for laser micro-spectrochemical analysis described on pp. 26, 42, 70, 83.

Laser micro-spectrochemical analysis, first suggested in 1962, has various advantages to offer over micro-spectrochemical analysis with contact electrodes: vaporization and excitation are separated in space and in time; there are no segregation processes as they occur in the continuous arc. Even with carbon electrodes, which are preferred for the micro-method II, and with prism spectrographs of medium dispersion no sections of the spectrum have to be disregarded in the interpretation because of band emission and resulting interference.

In the spectrochemical analysis of non-conducting samples with carbon or graphite electrodes, commonly practiced before, many cyanogen molecular bands are, as is well known, recorded in the wavelength range from 350 to 460 nm. Therefore in these cases the best possible advantage cannot usually be taken of the whole wavelength coverage of medium prism spectrographs. Different from spectra obtained with continuous and intermittent arcs and their usual exposure times from 10 seconds to several minutes, the spectra produced with the laser probe in the photographic emulsion in less than a micro-second are remarkable for their extraordinary sharpness.

Every analytical spectrum shows various carbon lines, for example C I 247·8 nm, C II 426·7 nm, 250·9 nm, 251·2 nm and C III 229·7 nm (Fig. 63 and 64), but not the whole gamut of interfering cyanogen band

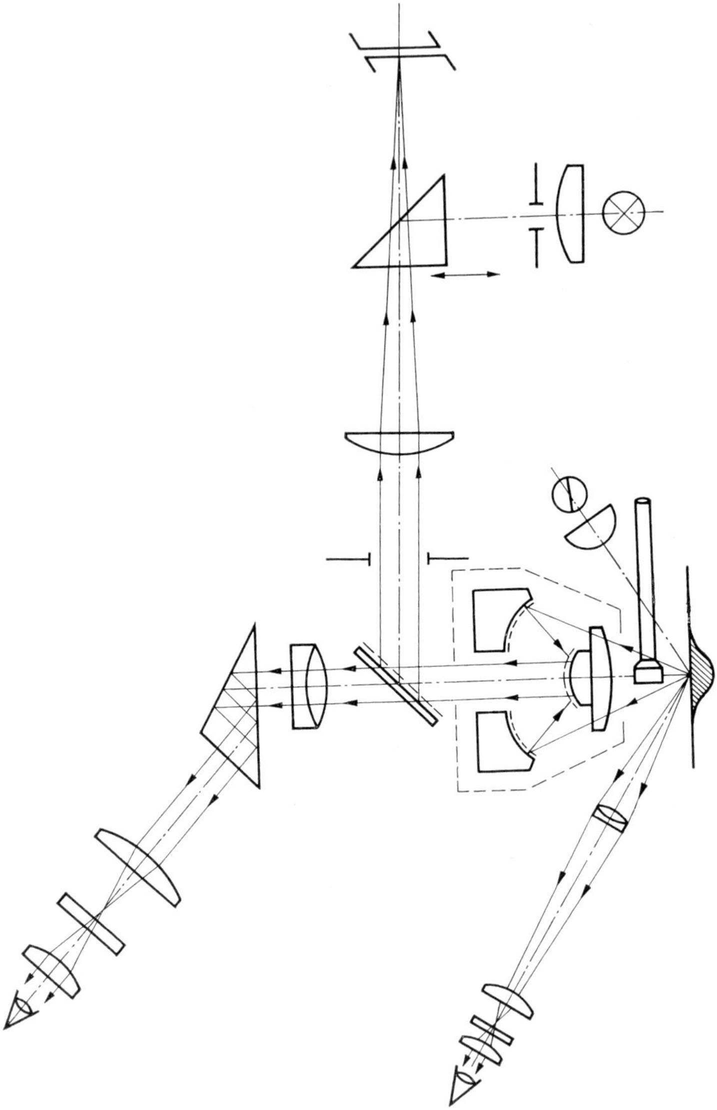

FIG. 62. Ray diagram of outfit for conventional spot analysis with contact electrodes (according to Koroljew[63b]).

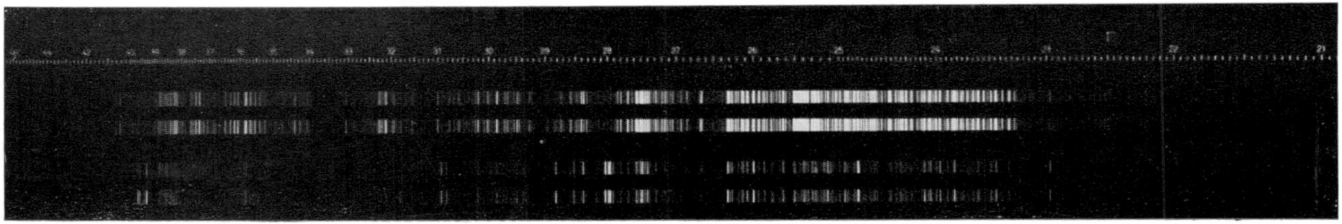

FIG. 63. Micro-analysis with different objectives (spectrograph Q 24).

From top to bottom: Pyrite reflecting objective ×40/0·5
flat-field achromat ×16/0·2
Almandine reflecting objective ×40/0·5
flat-field achromat ×16/0·2

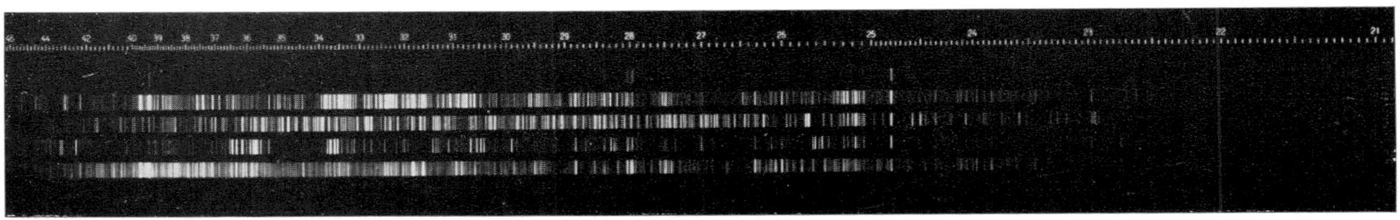

FIG. 64. Silicate analysis with the laser micro-spectrum analyser LMA fitted with the flat-field achromatic microscope objective ×16/0·2. From top to bottom: Titanite, Zircon, Thorveitite, Orthite

systems. If the electrode tips are correctly adjusted so that they are not imaged on the slit, only one weak cyanogen band between 380 and 389 nm appears. In practice this does not complicate the determination of any element and the entire wavelength range of a spectrograph of the type of the Q 24 can be utilized in the spectrochemical analysis of conducting or dielectric materials. Because of the short-duration impact of the laser radiation all elements are almost simultaneously vaporized. Samples of any shape, without any preparation, can be examined in the same way as the usual polished sections and thin slices.

The first qualitative micro-analysis of minerals was carried out by the head of the Analytical Chemistry Section, Petrological Sciences Division of the Geological Survey of Canada; the Jarrell-Ash Co. put their demonstration laboratory at the disposal of Maxwell[74] for test experiments with a laser microprobe outfit. First satisfactory results were only obtained on polished sections of ore minerals such as galena, chalcopyrite, sphalerite, bornite, niccolite, pyrite and chromite. The trace elements nickel (in pyrite) and silver (in galena) could be determined in addition to the major constituents.

Maxwell was able to detect the difference in chemical composition of zoned felspars and the homogeneity of other rock-forming minerals through point-by-point scanning with a series of consecutive laser pulses. Tests on polished thin slices of pyroxenite, granite, anorthosite and gabbro, however, were not quite satisfactory. An increase in energy of the Jarrell-Ash microprobe was called for. Maxwell said: '... may lead to the development of a laser beam of sufficient intensity to make the use of the auxiliary spark source unnecessary, thus improving the quality of the spectra obtained ...' In view of the fact that Brech[17] introduced a Q-switched ruby laser with a pulse power of 2 to 3 MW into scientific instrumentation, such a suggestion coming from a geologist must raise doubts

TABLE 13. Qualitative analysis of silicate minerals by laser micro-emission spectroscopy. (Reported by J. A. Maxwell, *Canadian Mineralogist*, 1963, *7*, 734) (Excerpt)

| *Mineral* | *Constituents* | | | |
|---|---|---|---|---|
| | *Major* | *Minor* | *Trace* | *Faint Trace* |
| Clinopyroxene | Si | Ca, Mg, Fe | Al | Ti |
| Orthopyroxene | Si | Mg, Ca, Fe | Mn | Ti, Al, Cr |
| Orthopyroxene | Si, Ca | Fe, Mg | — | Ti, Al, Cr |
| Plagioclase | — | Mg, Ca | Si, Al | B, Fe |
| | — | Ca | Mg, Si, Al | B, Fe |
| Potassium Felspar in | Si, Na | K, Al | Mg, Ca | Fe |
| metamorphic rock | Si, Na | K, Al | Ti, Mg, Ca | Mn, Fe |
| Potassium Felspar in granophyre | — | Si | Ca, Mg | Al |
| Enstatite (in polished section of meteorite) | Si, Ti | Mg | Fe, Al | — |

TABLE 14. Qualitative analysis of ore minerals by laser micro-emission spectroscopy. (Reported by J. A. Maxwell, *Canadian Mineralogist*, 1963, *7*, 731)

| *Mineral* | *Constituents* | | | |
|---|---|---|---|---|
| | *Major* | *Minor* | *Trace* | *Faint Trace* |
| Bornite | Fe, Cu | Cr | – | Mg, Si, Ca, Ag |
| Sphalerite | Zn | Fe | ±Si, Pb, Cd(?) | Cu, Si, Mg, Pb, Ca |
| Chalcopyrite | Fe, ±Cu | ±Cu, Al | ±Cr, Mg | Mg, Ca |
| Pyrite | Fe | Ni | – | Mg, Ca, Cu, Co, Ag |
| Pyrrhotite | Fe | Ni | – | Mg, Cr, Ca, Ag |
| Galena | Pb | – | Fe, Ag | Si, Mg, Bi, Ca |
| Siegenite | Ni, Fe | ±Cu, Zn | Co, Mg ±Cu, Zn | – |
| Niccolite | As | Ni | Mg, Fe | – |
| Skutterudite | As, Co | Ca, Fe | Ni, Cu, Cr | – |
| Arsenopyrite | Fe, As | Ni, Cr | Mg | – |
| Magnetite | Fe | – | Mg | Si, Mn, Ca |
| Chromite | Fe, Cr | ±Ti, Al, Mn, V | Mg, ±Al | ±Mn, Cu, Ca, V, Al |

± : Differences in composition can be found from sample to sample and in different parts of the same sample.

about the feasibility, unless macro-analysis or semi-micro-analysis be considered.

Maxwell obtained with the Q-switched ruby laser approximately semi-spherical craters of a diameter of about 100 μm and less, down to 50 μm. This shows that also with Q-switched lasers the spatial extent of the effect depends on the chemical nature of the sample and that the demand for further investigation into the influence of surface smoothness, hardness, thermal conductivity and crystallographic orientation on the reproducibility of the laser probe action is at least partly justified.

A further problem of the method is the provision of suitable reference standards for semi-quantitative micro-analysis.

In conclusion the Canadian scientist summed up his investigations approvingly: 'This is a very attractive method and, in the case of geological studies, should prove very useful in the future'.

In May 1963 Maxwell submitted another contribution to the *Canadian Mineralogist*[75] in which he reported on the progress of his work on polished ore sections and polished thin rock slices carried out at the invitation of the Jarrell-Ash Co. The identified major and minor constituents and trace elements of different minerals are listed in the Tables 13 and 14.

The authors conducted similar qualitative tests[13] and came to the conclusion that the femic rock mineral components are more satisfactorily vaporized by focused laser radiation than the bright salic components rich in $SiO_2$. It is furthermore beyond doubt that the total content of Fe, Mn and Mg of a mineral or its major constituents also has great influence on the reproducibility of a true micro-emission spectrum.

It is quite possible also to determine silicon by definite arc lines in the very spark-like light source, if for instance under a granule of plagioclase lies a mineral rich in iron and manganese which enhances the absorption of the red or infra-red laser radiation.

Table 14 indicates which chemical elements can be qualitatively determined in typical sulphide, arsenide and oxide ores.

A laser probe can similarly be employed for finding certain structure elements of meteorites. Apart from Fe and the minor constituents Ni and Si, the presence of the trace elements Cr, Mg and Ca can be established in iron meteorites, especially in the so-called kamacite.

In 1968, at Varna, Petrakiew *et al.*[98] reported on their analyses of the meteorite 'Pawel' which fell in the district of Tirnovo in 1966. The instrument used was the LMA 1 laser micro-spectrum analyser of VEB Carl Zeiss, Jena. Having had to deal with a stone meteorite they commented favourably on the advantage of laser micro-analysis of permitting direct testing of non-conducting samples without any difficulty; the availability of a polarization device for the LMA 1 model and the possibility of working with either transmitted and reflected light are mentioned as specially valuable features of the instrument. Table 15, taken from their work, summarizes the semi-quantitative results of the analysis.

TABLE 15. Results of semi-quantitative spectrochemical analysis of meteorite 'Pawel'[98]

| *Element* | *Percentage concentration* | | |
|---|---|---|---|
| | *Chondrules* | *Matrix* | *Crust* |
| Al | less than $1 \times 10^{-3}$ | approx. $3 \times 10^{-2}$ | approx. $5 \times 10^{-2}$ |
| Fe | major constituent | minor constituent | major constituent |
| Ca | approx. $1 \times 10^{-3}$ | 1 | approx. $3 \times 10^{-1}$ |
| Co | more than $1 \times 10^{-3}$ | – | approx. $3 \times 10^{-3}$ |
| Si | approx. $5 \times 10^{-3}$ | $3 \times 10^{-1}$ to 1 | approx. $1 \times 10^{-2}$ |
| Mg | approx. $5 \times 10^{-2}$ | more than 3 | 1 to 3 |
| Mn | approx. $1 \times 10^{-3}$ | less than $3 \times 10^{-3}$ | approx. $1 \times 10^{-2}$ |
| Cu | approx. $1 \times 10^{-3}$ | approx. $3 \times 10^{-3}$ | less than $1 \times 10^{-2}$ |
| Na | approx. $1 \times 10^{-3}$ | $1 \times 10^{-2}$ | approx. $1 \times 10^{-1}$ |
| Ni | approx. $3 \times 10^{-3}$ | approx. $1 \times 10^{-3}$ | $1 \times 10^{-2}$ to $1 \times 10^{-1}$ |
| Ti | – | approx. $1 \times 10^{-2}$ | approx. $1 \times 10^{-2}$ |
| Cr | approx. $1 \times 10^{-3}$ | approx. $1 \times 10^{-2}$ | approx. $1 \times 10^{-2}$ |

Although the microscope of the American laser microprobe only has illumination for work with reflected light, experiments were also carried out by irradiating thin polished sample slices with red laser radiation; simultaneous control by visual microscope observation in transmitted light, however, was not possible. Maxwell observed on these occasions that the very intense radiation produced relatively deep craters. Slices

30 μm thick were burnt through and the matrix material and part of the sample stage were vaporized. It follows from these experiments that thicker slices of about 50 μm have to be used as in work with electron probes (Geoscan of Cambridge Scientific Instrument Co., Cambridge).

The standard Zeiss LMA 1 outfit has a laser without Q-switch, emitting an energy of 0·1 to 1·0 J per pulse. A desired crater depth can be obtained through appropriately adjusting the supply unit parameters and choosing the appropriate diaphragms and objectives (see p. 107). The authors even succeeded in analysing thin layers and films considerably thinner than the usual polished sample slices: the various layers of a barrier layer photo-element, for instance, were examined one after the other to find the main constituents and trace elements.

The model LMA 1 is fitted with condensers for work with reflected and transmitted light so that a satisfactory microscopic examination—if required also with polarized light—of polished sections and of polished thin sample slices can always be undertaken.

The constituent elements of frequently occurring and important minerals, present in different concentrations, that were identified by Maxwell with single or repeated laser excitation (Tables 13 and 14) were also determined by the authors. In the analysis of galena with the laser micro-spectrum analyser (Fig. 65) the ultra-violet Pb I lines 283·3 and 280·2 nm could always be used for identifying the cation, furthermore the main analytical Pb line 405·78 nm, which otherwise is obscured by cyanogen bands, and then the typical set of lines 357·27, 363·96, 367·15, 368·35 and 373·99 nm. When recorded on 'blue extra hard' spectral plates the series of Pb lines that can be utilized for the analysis extends to the diffuse Pb II line 220·35 nm.

In galena from the Freiberg deposits the trace elements Ag and Cu can be determined without difficulty. A single laser pulse is sufficient, even with the entrance slit of the spectrograph model Q 24 set to a width of only 10 to 15 μm. The ultra-violet Ag I lines 328·07 and 338·28 nm and Cu I 327·4 and 324·75 nm served for identification.

The spectra were obtained in the two-step process of sample vaporization through pulsed laser radiation with coupled excitation in the electric spark gap between carbon electrodes inclined to the sample surface (cross excitation); the authors could always interpret these spectra with the help of the analytical lines of the chemical elements that were successfully used in earlier mineralogical investigations.[79] Hence the presence of elements can be established qualitatively in the usual way by means of analytical lines marked in the Q 24 tables of the iron spectrum.[129] Steel rods or iron carbonyl samples can be fitted in the electrode holder of the Zeiss microscope for laser spectrochemical analysis to produce a reference iron spectrum, but it is sufficient to incorporate systematically in the work programme for a spectrum plate some ferrous compounds, for instance Fe–Mn boracite or pyrite ($FeS_2$).

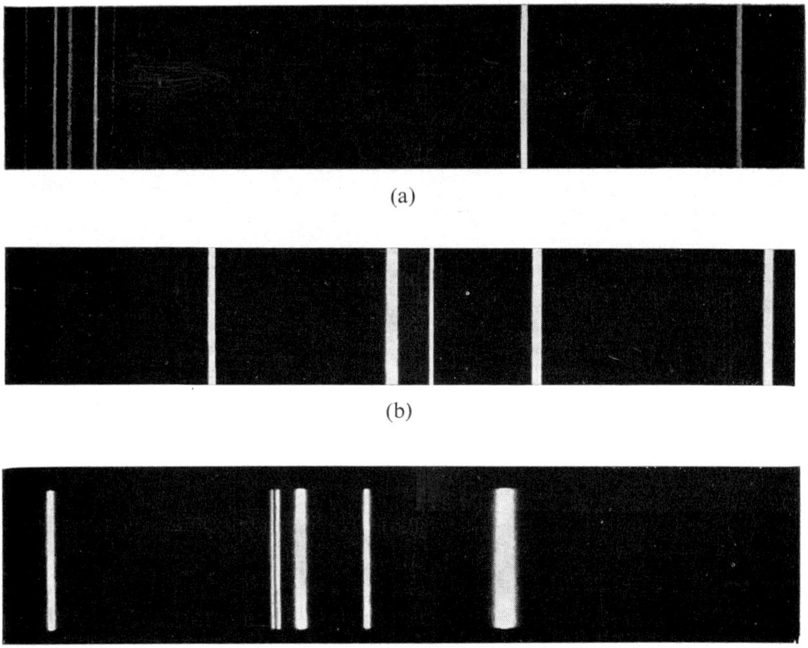

(a)

(b)

(c)

FIG. 65. Lead determination with laser microprobe.

(*a*) Analysis of crocoite from Tasmania. No interference by CN bands with the main analytical line 405·78 nm. On left: group of Cr I lines 425·4, 427·48, 428·97 nm.
(*b*) Typical group of Pb I lines in galena spectrum: 374, 368·35, 367·15, 363·96, 357·27 nm.
(*c*) Ultra-violet Pb I lines near the C II lines 283·67 and 263·76 nm, namely: 280·2, 282·3, 283·3, 287·3 nm.

In the examination of spectra with few lines a slit width of 25 μm is not objectionable. On the Jarrell-Ash laser microprobe, test spectrograms were even taken, evidently for intensity reasons, with a 35 μm slit.[55] When a sufficient quantity of the sample material is available, the authors, in routine examinations, use a slit width of 15 μm and produce with two laser flashes the vapour cloud necessary for synchronization of the capacitor discharge and for the spectrochemical detection of the elements present.

Similar to the examination of the sulphides mentioned before, namely those of Pb, Zn, Sb and Fe, the analysis of other sulphides did not cause any difficulties; amongst them were molybdenite $MoS_2$, cinnabar HgS, germanite $Cu_3(Fe, Ge)\ S_4$, pyrargyrite $Ag_3SbS_3$, proustite $Ag_3AsS_3$ and chalcopyrite $CuFeS_2$. And it was not necessary to operate the 0·5 J laser with full energy to determine the cations. Even very volatile elements like

Hg and As could easily be identified by their ultra-violet lines. Likewise successful was the attempt to excite sulphides in powder form placed loosely on a suitable support on the sample stage: in this way germanium and gallium could unmistakably be detected in germanite in the presence of copper and iron, apart from other determinations. The application of the Zeiss laser micro-spectrum analyser in a modified procedure to the analysis of sulphide and oxide ore components in polished sample sections too can decidedly be designated a rapid analysis; among the substances in question are chromite $FeCr_2O_4$, magnetite $Fe_3O_4$, hematite $Fe_2O_3$, ilmenite $FeTiO_3$ (Fig. 88) and rutile $TiO_2$.

With a slit width of 15 to 25 μm one laser pulse is sufficient for the paragenesis PbS, $CuFeS_2$, $FeS_2$; that means that the spectrum can be recorded in about one minute. And the sample is preserved as a record. The vaporized sample quantity varies, depending on the laser parameters and on the physical constants of the sample, especially its thermal conductivity, between 1 and 20 μg.

Experiments on cleavage planes of galena showed that in the {100} plane craters can form either in the shape of circular cones of, for instance, 50 μm radius and 115 μm depth, or, with a different a laser energy, as approximately hemispherical indentations of perhaps 70 μm radius. In this case the vaporized sample quantity is 2·3 μg PbS; which means that the limit of detection for Ag is less than $2 \times 10^{-8}$ g. If, however, the same sulphide sample is excited by a continuous arc in evaporation from an auxiliary carbon electrode, the limit of detection for Ag is $5 \times 10^{-7}$ g. This improvement of the absolute limit of detection is achieved without resorting to any special working procedure.

Subsequent investigations have shown that, in the analysis of galena and pyrite with the LMA 1 model, application of a magnetic field can still further enhance the possibility of detection.[40]

The points of impact of the laser beam in sulphides are often surrounded by a fusion ridge (Fig. 66). Under the microscope the irradiated areas look like small volcanoes. The thermal effect frequently causes new cleavages and the sample surface has a tempering tint which can be clearly seen in the reflected bright field.

Some samples in a characteristic manner tend to eject fusion beads, as for instance the arsenate erythrite illustrated in Fig. 52. On silicates, borates and carbonates (Fig. 67) impact areas with more or less smooth edges can often be seen.

When irradiating plagioclase with a giant pulse ruby laser one obtains, according to Maxwell, only irregular craters with a fusion region of milky appearance. The action of Q-switched solid-state lasers on certain silicates causes, in addition to the very irregular shape of the affected area, spattering of fused material over the sample surface. According to the experimental results so far known, lasers without Q-switch are very suitable for the analysis of silicates even if the samples have not been prepared by polishing a plane section surface or by cutting out and polishing a thin slice.

The same laser of a given performance always vaporizes smaller quantities of silicate than of sulphide samples.

Mohr[88] tested the LMA 1 model and its laser without Q-switch and with passive Q-switch on different silicates, namely sanidine, microcline,

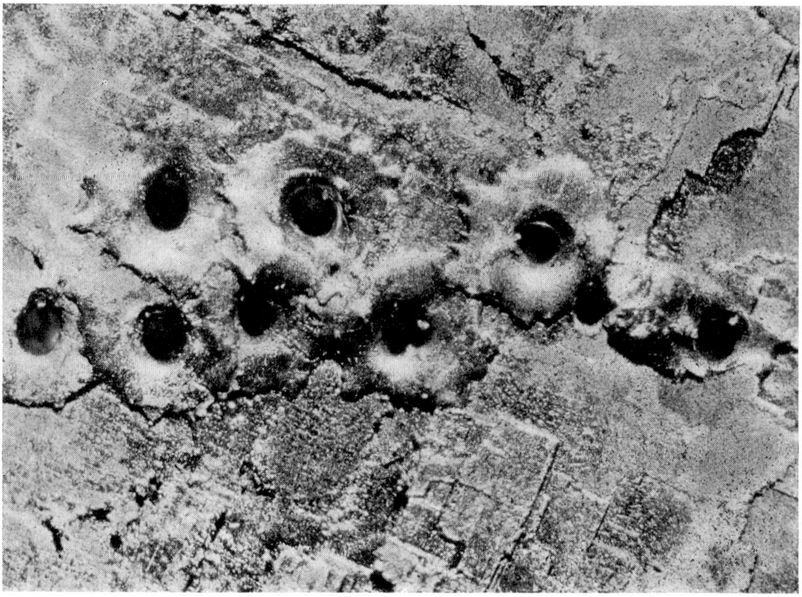

FIG. 66. {100} surface of galena crystal with effects of several laser pulse impacts showing fusion ridges.

Wavelength: 1060 nm. Crater diamaters: 140 μm.

heulandite, scolecite, montmorillonite, indianite and pollucite. Besides the major constituents of pollucite, which are Si, Al, K, Ca and Cs, the following elements could be detected: Ba, Na, Mn, Fe and Ti. No improvement of the sensitivity resulted from vaporization with a laser with Q-switch compared with a non-switched laser. The conditions are different for transparent quartz and beryl: Ti, Fe, Al, Cu, Na, Ca and Ba could be determined in quartz only with a Q-switched laser. In beryl Ca, Mg and Fe have already been found without auxiliary spark excitation.

The use of resonators of different materials affects crater dimensions. The solid-state laser model ZFL 750 fitted with a ruby rod produced a crater of 60 μm depth and 40 μm radius in an olivine crystal [$(Mg, Fe)_2 SiO_4$].

With a neodymium-doped glass resonator pumped by the same helical xenon flash tube of the same energy, the crater depth and radius were 120 μm and 25 μm, respectively. In these cases sample quantities of about 0·3 μg were vaporized.

Karjakin, Achmanova and Kaigorodov[59] worked, amongst other projects, on the quantitative determination of Si, Mg and Fe in silicate minerals.

The authors investigated many anisodesmic and mesodesmic substances from the big group of crystals with structures which are distinguished by resonance bonds between covalent and electrocovalent types, so-called UP resonance lattices. In order to extend the field of application of the Zeiss outfit for laser micro-spectroscopy to as many groups of elements of the periodic system as possible, cations of different components

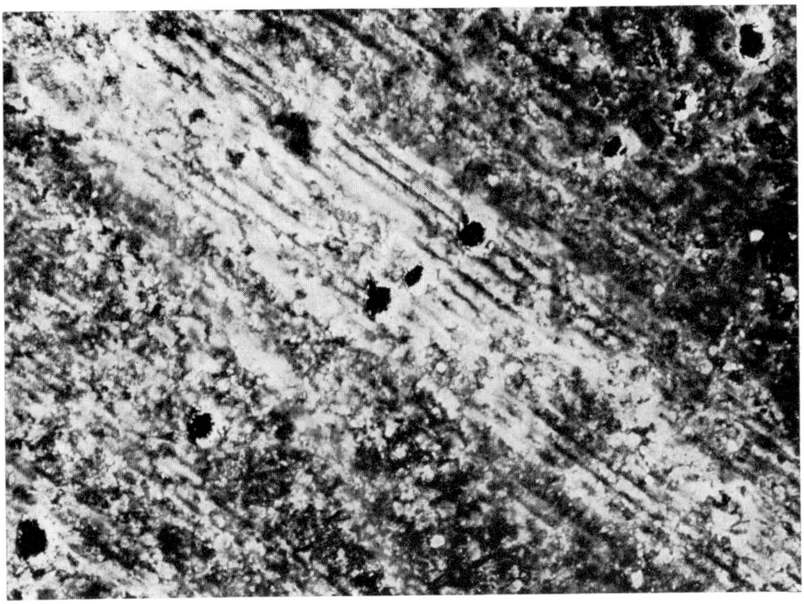

FIG. 67. Effects of impact of ruby laser radiation (wavelength 694·3 nm) on a copper carbonate sample (malachite). Average crater diameter 40 μm.

were determined in the ultra-violet and in the visible spectrum region, namely of carbonates–Ca in calcite, Ba in witherite, Mn in rhodochrosite, Cu in malachite and azurite; of some sulphates–Sr in celestite, Ca in anhydrite and in gypsum; and of several borates–Fe, Mn, Mg besides B in boracites (Fig. 68).

Determinations of constituent elements of phosphates included Cu and Al in turquoise; Ca and U in the uran mica autunite; Li, besides Fe and Mn, in triphylite, using the line 323·2 nm. Cr and Pb lines are predominant in the spectrum of the chromate mineral crocoite. W, Fe and Mn lines were always found in the spectra of mixed crystals of the wolframite series and, therefore, tungstates can also be examined with the laser microprobe (Fig. 88). The arsenates erythrite, annabergite and cabrerite served as targets for the determination of Co, Ni and As in these compounds with mesmodesmic tendency (Fig. 88). Vanadates too, in the form of minute pucherite crystals, were found to be suitable test materials in which Bi and V could satisfactorily be detected (Fig. 88).

FIG. 68. Section of boracite spectrum from 249 to 261 nm
The marked spectrum lines are sufficient for definite identification of the material. From right to left: B I 249·68, 249·77 (C II 250·9, 251·17/251·2); Mn II 257·6, 259·37; Fe II 259·84, 259·94; Mn II 260·57 nm.

Si and Al could always be determined from the anion structure of silicates with different linkage of the $SiO_4$ tetrahedra. Numerous cations could also be identified, such as Be in the nesosilicate phenakite, Mg and Fe in the nesosilicate olivine (Fig. 79), Zr in the nesosilicate zircon, Ca and Ti in the neso-subsilicate titanite, Th in the nesosilicate thorite, Zn in the nesosilicate willemite, Sc in the sorosilicate thortveitite, Mg, Fe, Mn, Ca, Al and Cr in the mixed crystals of the garnet group. Also trace elements were successfully established in silicates that were irradiated and analysed in the form of crystal fragments, cleavage segments or of very small crystals grown in or on rock specimens (Fig. 69); amongst them were Mn in willemite, $Zn_2SiO_4$, Y and Th in the titanite contained in the syenite of the Plauen valley near Dresden, and Hf in zircon.

Particularly successful were the attempts to determine with an outfit consisting of laser microscope, auxiliary spark gap and Q 24 quartz prism spectrograph elements of the group of the rare earths which otherwise are mainly analysed with grating spectrographs.[86] It was possible to establish clearly the presence of La, Ce, Yb and Y in orthite (Fig. 76) and of Ce and Yb in monazite. Isodesmic substances, especially oxides, served for proving the possibility of detecting U in pitchblende and of Nb in samarskite (Fig. 84).

The outfit advisable for work on the alkali metals Li, Na, K, Rb and Cs, best in the wavelength range from 568 to 895 nm, should include a spectrograph with the three-prism dispersing system according to Försterling instead of the spectrograph model Q 24.

A great variety of chemical elements, listed in Table 16, among them Au, In, Tl, Sb, Cd, Ga, Ta, Pd, Rh, Te, were determined by vaporizing micro-quantities of the corresponding metals which partly occur in nature as minerals, partly in alloys and chemicals.

The laser micro-spectrum analyser, therefore, can also be used to advantage in metallography and for testing materials (see § 4.1.2). A preparation of the sample, as otherwise usually required, is here not necessary; hence time is saved and the precision is increased because there is no risk of contamination, no loss of material, no error in weighing, etc.

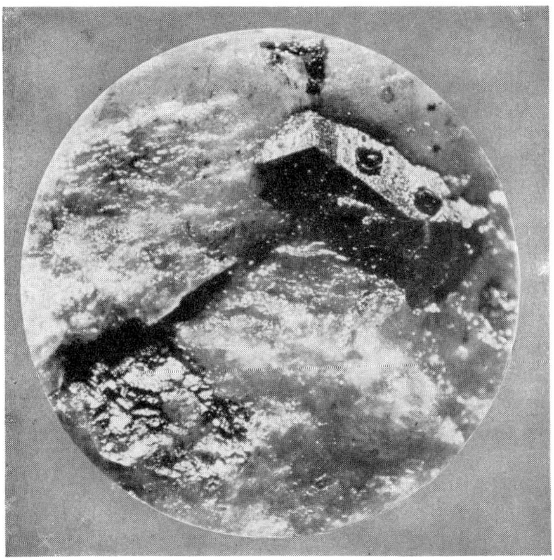

FIG. 69. Effect of impact of infra-red laser radiation ($\lambda$ = 1060 nm) on titanite crystal (nesosubsilicate $CaTi[O/SiO_4]$) grown in syenite rock from the Plauen district near Dresden.

So far the authors have succeeded in determining sixty elements in a variety of natural and manufactured inorganic materials (Table 16). With the new method the absolute limits of detection of some of these elements are considerably less than $10^{-6}$ g which is the order of magnitude of the vaporized sample quantity. For finding faint traces the vaporized quantity can be increased by raising the laser energy, by utilizing the principal analytical lines with repeated laser firings and by employing other microscope objectives, for instance the flat-field achromat $\times 16/0{\cdot}20$ of the LMA 1 instrument. With this refracting objective, crater diameters of 200 to 250 μm can be obtained, with resulting improvement of the relative limits of detection.

The LMA 1 laser micro-spectrum analyser has a special polarizing microscope for work with transmitted and reflected light. To the scientist anxious to establish, as far as possible, all measurable properties of minerals this outfit offers, besides the possibility of carrying out the usual microscopic examinations, simultaneous spectrochemical analysis as a further source of information. Petrography pursued for the sake of technical applications has somewhat different aims: basically identification of a mineral is sufficient. Every mineral has many more properties than are required for its identification; hence the scientist in the laboratory is in the fortunate position to be able to choose the properties most easily and quickly found. Rapid analysis of the chemical composition of an unknown substance coupled with simultaneous visual examination under the microscope usually suffices for identification. The experimental worker appreciates the saving of laborious sample preparation or of separation of mineral mixtures: that is, in the former case, of producing rock slices or preparing polished sections, and, in the latter case, of segregation under the stereo-microscope or fractional separation in often poisonous liquids of high specific gravity. An example is the determination of a hard mineral component in diatomite used for polishing. Isolating small crystals

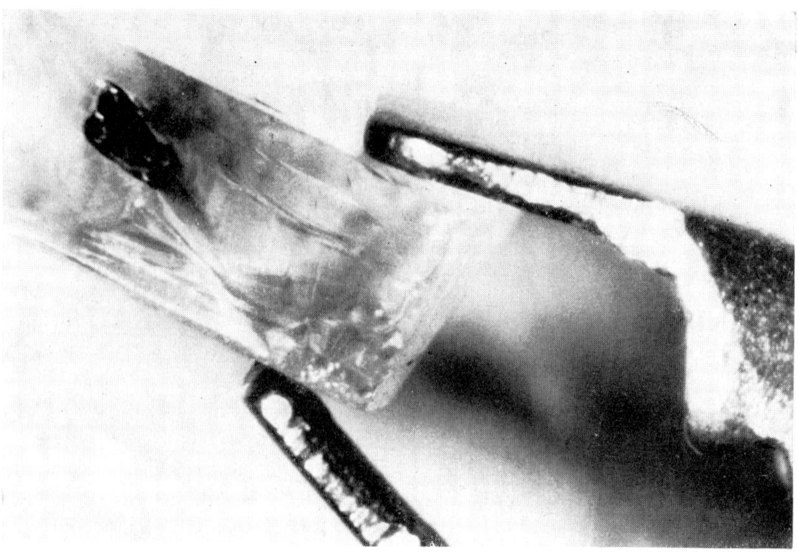

FIG. 70. Rutile inclusion in quartz.

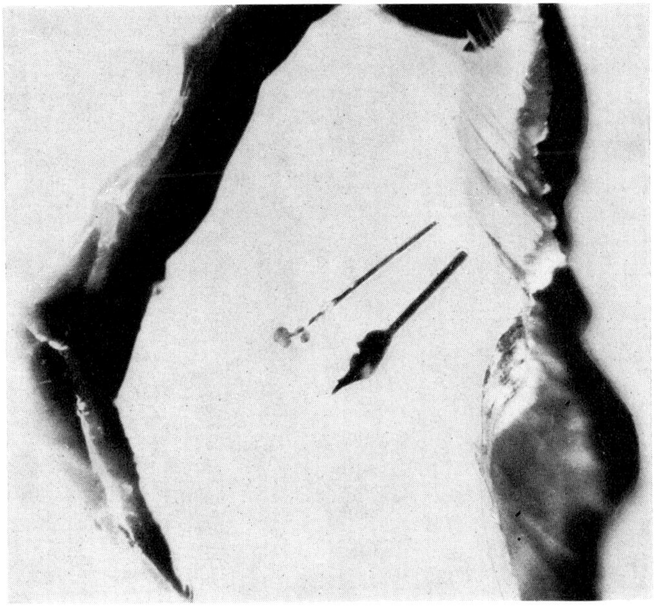

a)

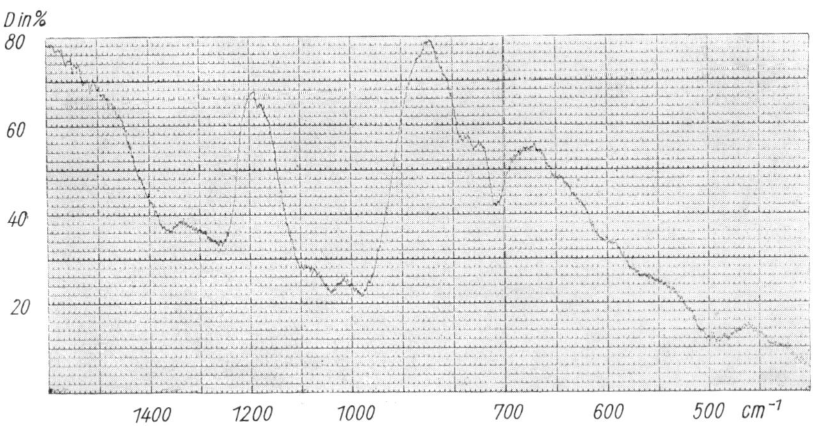


b)

FIG. 71. (*a*) Needle-shaped cyclosilicate inclusion (tourmaline) in quartz (Brazil) with effects of two infra-red laser radiation impacts (clearly visible in shadow). (*b*) UR-10 spectrophotometric trace of the needle-shaped tourmaline crystal.

(Fig. 70) is very often difficult, but it can conveniently be obviated by resorting to a micro-method. Once the authors had to analyse a needle-shaped crystal enclosed in a clear transparent mineral. A rutile needle, $TiO_2$, was conjectured. Laser micro-spectrochemical analysis, however, showed there was no Ti, but B, Mg and Fe were unmistakably identified in the sample. Hence the finding was: tourmaline needle. To confirm the result a micro-absorption analysis was subsequently carried out with a suitable attachment to the UR 10 spectrophotometer and the typical diagram of the cyclosilicate was obtained. A photograph of the crystal and its absorption spectrum taken on the UR 10 spectrophotometer are reproduced in Fig. 71(a) and (b).[16] This is an example of the successful combination of two modern techniques; further suggestions are to be found in Chapter 5. By this method of combining analytical techniques the authors found besides tourmaline and rutile also dolomite, sphalerite, ammonium sulphate and talc in Brazilian quartz.[16]

The spectroscopic examination of differently coloured pigment zones is of certain importance for the classification of agate samples. Fig. 72 shows the results of three laser impacts on brown pigment patches of an agate from South Africa; the produced spectra contain lines of Fe, Ca, Mg. In red pigments of the same sample only Fe and Mg were found. The craters had a diameter of 100 μm and a depth of 60 μm. The LMA 1 analyser was fitted with a neodymium glass resonator, already used for some time, and an objective ×16/0·2. The settings of the mains supply unit were: capacitance (for laser) 508 μF, flash tube voltage 2 kV; electrodes: voltage 1·5 kV, capacitance 2·5 μF, inductance 60 μH. The spectrograph was the Q 24 model with condensing optics of 80 mm focal length, used at an aperture ratio of *f*/15 and with the slit set to a width of 14 μm. These settings are generally suitable for examinations of minerals and rocks as described.

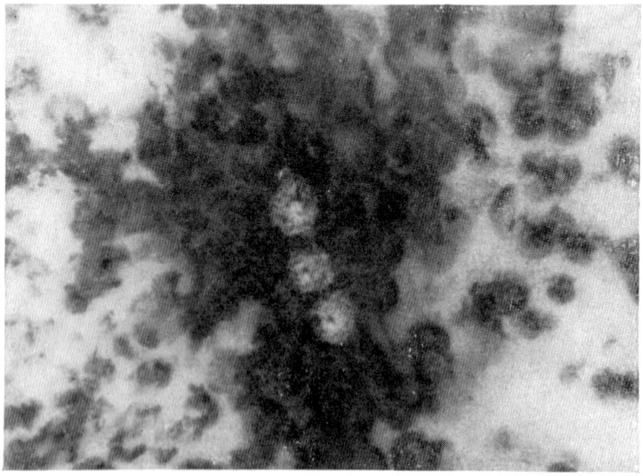

FIG. 72. Three laser craters in the pigment of an agate sample from Africa.

Knowledge of the chemical composition of a sample permits immediate recognition of pseudomorphs. Such minerals of secondary formation, which often retain the forms of the primary crystals, are of interest for discovering the natural history of a paragenesis and sometimes also for the science of mineral deposits.

Laser micro-spectrochemical analysis will be adopted as a method of investigation in geochemistry for such problems as the determination of key elements and minerals, the study of the geochemical distribution of elements and of the relative age of minerals.

Unfortunately no specific germanium ore deposits exist for producing germanium, today so important for the semiconductor technique. The supply depends on the low contents in sphalerite and in carbons or on finds of the mineral germanite.

Uranium is not only found in oxide ores of luminous colours, such an attraction for the prospecting geologist. Most of it comes from bituminous slates or carbonaceous matter similar to the gold conglomerates of the Witwatersrand in South Africa.

The following pages deal with the application of spectrochemical analysis to the science of mineral deposits and to carbon petrography.

*Ores and smelting preparation products*

Microscopy of ores was developed rather late as a special discipline of general microscopic methods in mineralogy.[81] The technique of microscopy by reflected light was known and established in the study of meteorites before the importance of examining serviceable ores was recognized. Only full knowledge of the mineral composition of an ore to be used for smelting makes it possible to say if the ore can be prepared, to decide on promising methods of preparation, to evaluate adopted procedures and to control their correct application. Previously microscopic and chemical examinations had to be carried out separately by mineralogists and experienced analytical chemists. If collaboration was not very close misunderstandings often arose leading to unsatisfactory results of one test or the other.

To be able to identify reliably all minerals and ores of a deposit one has to be fully versed in mineralogy, microscopy of ores, petrography and the science of mineral deposits; especially to have many years experience in the techniques of the microscopy of ores; and finally to have available an extensive collection of comparison material selected in anticipation of future needs. Specialized knowledge is furthermore required for the chemical and microchemical methods of operation to be applied. The mining and smelting industries therefore depend on assistance from university institutes or have to employ specialist graduates and to set up laboratories for these particular applications.

Micro-spectrochemical analysis with solid-state lasers in this respect serves two purposes. First, the new instrument links microscopy and chemical analysis—the simultaneous analysis simplifying or controlling the microscopy and confirming its results. Second, the work with the LMA outfit can also be undertaken by personnel trained in appropriate technical institutes and by laboratory technicians. Micro-spectroscopy can be carried out on unprepared fist-size mineral samples, on polished sections and thin polished slices of all dimensions, on granular specimens and odd fragments.

After identification of a mineral, the chemical composition of its individual constituents is first of all of interest for the preparation for smelting because this alone allows the possibilites of separation to be judged. Here the great information content of a spectrum is of special importance, for instance of a spectrum taken for preliminary inspection with a spectrograph of the Q 24 type. Such a spectrum not only indicates the presence or absence of elements looked for, but also reveals unexpected components which can be decisive for the success of a chemical preparation for smelting.

Significant for preparation techniques are isomorphous replacements or additions of particular elements that may be a reason for mining otherwise worthless minerals. Examples are the gold content of pyrites, nickel and cobalt content of iron sulphides, cobalt content of manganese ore dust etc.

Reference has still to be made to the manganese ores, one of the mineral groups most difficult to identify. Especially complicated is the microscopic examination because the optical properties of these mostly fine-grain, xenomorphic minerals overlap to a great extent; this is particularly so with the technically most important group of the sedimentary earthy or dust ores of manganese. Polished sections are practically impossible to produce and the sample preparation for electron microscopy is very time-consuming. The results of X-ray investigations of powder samples sometimes are ambiguous. Most of the manganese oxide ores of sedimentary origin were originally precipitated as gels and the subsequent crystallization was only more or less complete.

As the result of a laser micro-spectrochemical analysis carried out in a few seconds and showing the presence of certain elements one can, for instance, arrive at the following conclusions:

Ba content besides Mn indicates psilomelane;
Ba–Fe content besides Mn indicates hollandite;
Pb content besides Mn indicates coronadite or cesarolite;
Zn content besides Mn indicates hetaerolite;
Cu content besides Mn indicates crednerite.

Additives due to adsorption are the rule especially in minerals that have crystallized out of gels, as for instance melnikovite-pyrite or pyrolusite which can contain Pb, Cu, As, Co, Ni and other elements in varying quantities. All forms of chemical heterogeneity, usually combined with physical heterogeneity, are of importance for the mineral preparation. The different behaviour in flotation of some galena ores, for instance, is due to the presence of cerussite that has been formed in certain zones by oxidation.

The presence of a certain metal in a concentrate may be due to inclusions (copper in sphalerite, gold and platinum in sulphides and arsenides). The silver content of ores from galena deposits comes from tetrahedrite with high silver concentration, intergrowths with argentite, ruby silver minerals, native silver, stephanite, polybasite, etc. These sources of silver can often be hardly recognized under the microscope. Inclusions of tetrahedrite frequently accumulate on the grain boundaries of galena. They are liable to be rubbed off in intensive crushing and to get into the finest slurries in wet preparation and so to be lost. Of similar interest is the coupling of bismuth with galena.

The microscopic examination of gold ores and products of the gold ore preparation is difficult because of the minute quantities involved which, in fact, are often missed. 'Tarnished gold' is gold covered by a thin film of a material without gold content. This film usually is of sub-microscopic thickness, of the order of 1 $\mu$m. It is no problem to penetrate the covering layer with a laser firing for an analysis of the material beneath it. This is equally true for thicker layers such as those of 'rolled gold'. As a matter of fact most of the native gold is covered by such a film which consists of iron oxide ('rusty gold') or also of other materials. In practical work with laser micro-analysers on samples covered by some thick layer, solid-state lasers without Q-switch are preferable.

### *Application in coal petrography*

The solid fuels of fossil origin contain up to 40 per cent inorganic substances. For decades the investigations into the mineral contents of coal have been of greatest interest. The nature as well as the quantity of its mineral contents determine the value of coal in fuel technology both for burning and for coking. Knowledge of the inorganic components is of importance for a number of reasons: in coking because of possible noxious substances; in the chemistry of the coal on account of catalysts and contact poisons or because of possible new formations of minerals in the course of processing; in the preparation of the coal and in mineral extraction; in geological problems of seam formation and in stratigraphy; for mine safety aspects of silicosis, risks of seam fires, etc.

The most important of the inorganic substances that occur are: quartz, felspar, zircon, pyrite, siderite, chalcopyrite, pyrrhotite, hematite, clay minerals, calcite, ankerite, aragonite, dolomite, less frequently galena and sphalerite.

Micro-spectrochemical analysis with laser excitation will become an important addition to the existing methods of examination and will lead to increased accuracy in complete petrographic coal analyses.

It is known that certain trace elements (Zn, V, Ti, Ge, Ni, Co, Mo, Cr and Cu) are selectively stored in plants so that in coal their occurrence is confined to certain veins. Finding the distribution and concentration of these elements is an interesting field of application of laser microprobe work.

In 1968 scientists co-operating with CSIRO (Commonwealth Scientific and Industrial Research Organization), Division of Mineral Chemistry, Sydney, reported that Ti, Mg, Be, Si, Al, Fe, Na, Ba and Sr in particular could be determined with great sensitivity in coal samples from a mineral matrix with a high content of Ca and P. It was possible to find the required Ca : Mg ratio of dolomites from coal samples by using standard reference pressed disks containing different concentrations of $CaCo_3$ and $MgCO_3$ powders.[15]

## 4.1.2. Metallography

The following interesting remarks are made in the introduction to *Metallography* by Schumann:[1176] 'Andererseits sind einer weitergehenden Gefügeinterpretation gewisse Grenzen gesetzt durch das Unvermögen, bestimmte Substanzen im Gefüge mikroskopisch zu erkennen oder zu identifizieren, ... Auch durch Einführung des Elektronenmikroskops, des polarisierten Lichtes, des Phasenkontrastverfahrens und verschiedener Spezialätzmittel in neuerer Zeit konnte diese Grenze vorläufig nur geringfügig verschoben werden. Neue Möglichkeiten eröffnet die in den letzten Jahren entwickelte Mikrosonde.' [On the other hand certain limits are set to a more penetrating structure interpretation by the fact that certain substances within the structure cannot be recognized or identified by microscopy, ... So far this limitation could only be slightly relaxed even by such developments as electron microscopy or the use of polarized light, the phase contrast method and various special etching agents. New possibilities have been offered by the microprobe developed during the last few years.]

In Chapter 5 the advantages and disadvantages of the electron microprobe are discussed and it is shown how the newly introduced laser microspectroscopy fills a still existing gap.

The laser method offers two special modes of application:

1 Chemical analysis of areas of unprepared samples or their polished sections on a microscopic scale.

2 Analysis of macroscopic samples in a non-destructive test, using up material of about 1 $\mu$g.

To deal first with unprepared samples, before the analysis these have to be cleaned of all traces of oil, paint, varnish, etc. Work on such samples is restricted by their size. On the first Zeiss LMA model a maximum height of 8 cm can be accommodated whereas the lateral dimensions are not limited. In the further development of the instrument, however, the model LMA 1 as a safety measure had to be provided with protecting doors resulting in a maximum acceptable sample size of 100 × 100 × 75 mm (Fig. 73).

It is sometimes advisable to embed small samples in a matrix, for example synthetic resin, pizein, sealing wax, gypsum, but not, as usual in metal microscopy, in low-melting-point alloys such as wood metal, an alloy of Bi, Pb, Sn, Cd with a melting point of 60 °C; the vaporization of these volatile elements can distort the result of the chemical analysis. The

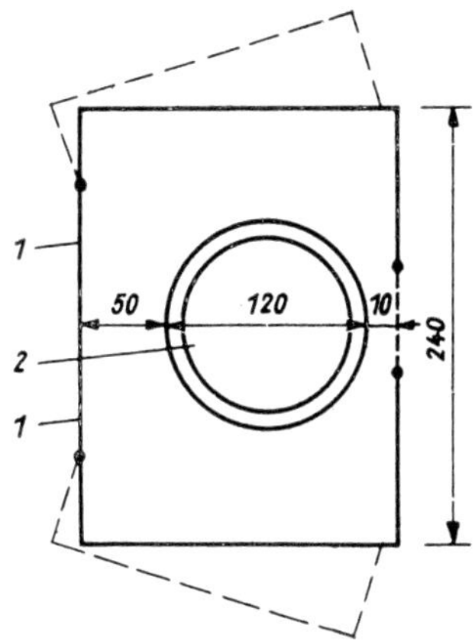


FIG. 73. Sample space in the LMA 1 model.
1 Safety doors, closed; 2 Microscope stage.

chemical composition of any other matrix material also has to be ascertained in a control analysis to prevent interferences and wrong conclusions.

To prepare a polished section for microscopical examination the sample, after being embedded in a matrix, has to be ground and polished. These two operations are not required for a laser micro-spectrochemical analysis if the sample region to be investigated is already visible. In very soft metals, such as lead and tin, the sample part of interest can often be exposed by means of a microtome or a sharp razor blade. For very hard alloys a guillotine has to be used. The sample must not be allowed to become too warm in these manipulations because structural or chemical changesm ight then occur.

Laser micro-analysis of unprepared samples is important for finding the cause of defects in fractures, pores and similar irregularties.

It is well known in metallography that a flat, very smooth surface is the most important prerequisite for obtaining a good microscopic image of metallic materials. Also in laser micro-spectrochemical analysis the examination of polished sections will always be of interest. In rolled sheet metal and strips, for instance, only consideration of the texture of the polished longitudinal, transverse and horizontal sections gives a general picture of the spatial distribution of the analysed heterogeneous matter.

In many cases the necessity of a micro-analysis only becomes apparent after preparation and polishing of the sample in the course of the examination under the microscope, because frequently the structure components of an alloy only show up in the polished surface (Fig. 71).

Fundamentally there is no objection to the use of etching, but the following has to be kept in mind: the chemical composition of the etching liquid and its impurities must be exactly known. All traces of the etching liquid and of the reaction products must be completely removed after the etching. Finally the particular mode of action of every etching reagent must be known.

Grain boundary etching causes a dissolution of the layers (more or less thick) of metallic or non-metallic structure components or impurities to be found at the grain boundaries. These extracted components can, therefore, no longer serve as targets for the spectrochemical analysis by the laser microprobe. Formation of an oxide layer following grain surface etching will normally not interfere with this analysis. A grain colouring effect can, however, also be produced if certain components are dissolved out of the alloy itself. When, for instance, duralumin, an aluminium alloy containing 4 per cent copper, is etched with $HCl + HF + HNO_3$, the copper is dissolved out and then deposited as a film on certain crystal planes, namely $\{100\}$ cube faces, but not on $\{111\}$ octahedron faces. In certain cases it is therefore advisable to etch only one half of the polished surface.

The following structures are investigated in metallography: metal crystallites, mixed crystals, intermetallic phases, intergranular extrusions, segregations, slags; furthermore, diffusion and recrystallization processes, soldered joints, welding seams.

An important problem in cases of damage is that of finding the cause. The test objects here are, for instance, segregation products, gas bubbles and crystals formed in the cavity, endogenous inclusions (Fig. 75) and exogeneous inclusions such as impurities that have entered the melt, for example from the mould sand, stirring pole, incompletely fused additions, etc.

A test that was successfully carried out with a laser probe on the rotor of a jet engine[51] concerned a subject lying between metallurgy and forensic science: it was suspected that a part of a special alloy had been treated by welding which was not permissible. A localized concentration of boron was detected in the area in question; it had resulted from the use of a flux containing boron. This element that gave the clue could not be detected by X-ray analysis.

Furthermore all raw materials and ores (see § 4.1.1.), as well as the products of ore preparation for smelting and furnace slags, can be investigated with the laser analyser.

*Ferrous alloys—steels*

The following elements can be detected in alloy steels by means of laser spectrochemical analysis: Fe, Mn, Si, Al, Ni, Cr, V, Mo, W, Co, Ti, Ta, Zr. It is not possible to determine H, O, S, P and, if carbon or graphite auxiliary electrodes are used, C.

Volkmann[131] analysing steel samples with the model LMA 1 (plane grating spectrograph PGS 2, discharge imaged directly on the spectrograph slit, vaporized sample quantities between $0{\cdot}5 \times 10^{-7}$ and $11 \times 10^{-7}$ g) established the following limits of detection:

| | | | |
|---|---|---|---|
| Cu: | $5 \times 10^{-11}$ g | V: | $7{\cdot}5 \times 10^{-10}$ g |
| Mn: | $6 \times 10^{-11}$ g | Co: | $1 \times 10^{-9}$ g |
| Cr: | $1{\cdot}5 \times 10^{-10}$ g | Al: | $2{\cdot}5 \times 10^{-9}$ g |
| Si, Ni: | $3 \times 10^{-10}$ g | Mo: | $4 \times 10^{-9}$ g |
| Ti: | $5 \times 10^{-10}$ g | | |

For determining steel components with the outfit consisting of the Q 24 spectrograph, microscope LM and laser ZFL 1000, one can use the following analytical lines:

| | |
|---|---|
| Cr: | 425·4 and 357·9 nm |
| Cu: | 327·4 and 324·8 nm |
| Mn: | 257·6 nm |
| Si: | 288·1, 252·8 and quintuplet 252·4, 251·9, 251·6, 251·4, 250·7 nm |
| Mo: | 287·1, 284·8, 281·6, 277·5 nm |
| Ni: | 325·4 and 341·4 nm (interference by Fe) |
| V: | Triplet 318·5, 318·4, 318·3 nm and 309·3 nm |
| C: | 229·7 nm |

Important for semi-quantitative and qualitative steel analyses is the report that tensile straining in special steels causes an intensity increase in chromium analytical lines.[25] Such matrix effects apparently affect the sample quantity that is being vaporized when lasers with comparatively low output energy (about 0·2 J) are used.

Pure metal electrodes, for instance of Mg, 50 mm long and of 5 mm diameter, with a 10° tip, serve for determining carbon. The described outfit is however not very suitable for the routine analysis of carbon in steel. For such work, as well as for the analysis of S and P, the electrode holder has to be mounted in a vacuum chamber and the laser micro-analyser must be linked with a vacuum spectrograph.

Every steel contains certain quantities, large or small, of non-metallic constituents, so-called slags. Endogenous slags are mostly compounds of FeO, $Fe_2O_3$, FeS, $SiO_2$, MnO, MnS, $Al_2O_3$, CaO, TiN, ZrN, $Cr_2O_3$. Very often one finds silicates and oxides such as fayalite, spinel, chromite etc.

'Die metallographische Untersuchung der nichtmetallischen Einschlüsse ist sehr schwierig und nur dem Geübten bis zu einem gewissen Grade möglich.' [The metallographic examination of the non-metallic inclusions is very difficult and only within the scope, to a degree, of an experienced worker.] (Schumann,[117] p. 355.)

Determinations in polished sections depend on many factors such as

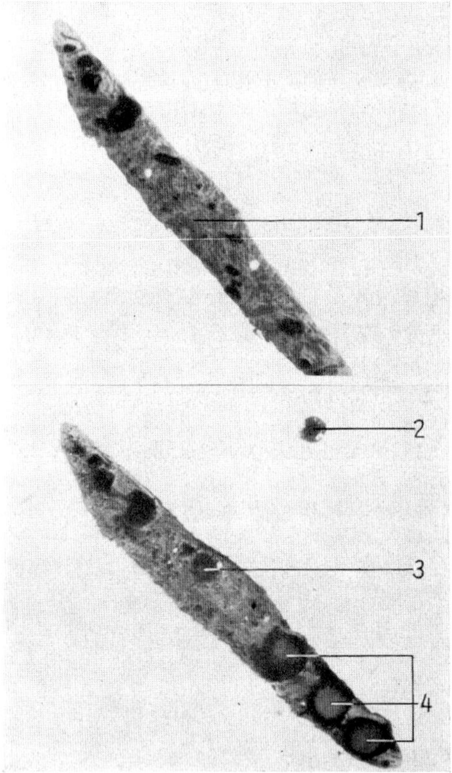


FIG. 74. Silicate inclusion in a steel sample. Photograph taken with the photomicrography attachment to the LMA 1 model.

1 Silicate inclusion before laser irradiation; 2 Laser crater of 25 μm diameter; 3 Laser crater of 25 μm diameter; 4 Three laser craters of 60 μm diameter.

size, shape, colour, hardness and photo-elasticity, and on the effects of heating and etching the polished surface. These time-consuming and not very reliable procedures are described in the literature for common compounds, but they usually fail completely in the examination of exogenous slags of the many different compositions.

Schumann[117] (p. 353) says:

'Eine genaue qualitative und quantitative Bestimmung der nichtmetallischen Einschlüsse im Stahl allein mit Hilfe metallographischer Methoden ist nicht möglich. Zu diesem Zweck muß man auf die rückstandsanalytischen Verfahren zurückgreifen. Dazu wird die metallische Grundmasse der zu untersuchenden Probe durch geeignete Säuren, durch Chlor oder durch Elektrolyse aufgelöst. Die im Rückstand befindlichen Carbide werden durch Oxydation zerstört, und die zurückbleibenden nichtmetallischen Einschlüsse werden nach den Methoden der Mikroanalyse untersucht.' [Exact qualitative and quantitative determination of the non-metallic inclusions in steel is not possible by metallographic methods alone. For this purpose one has to resort to residue analysis. Here the basic metal material of the test sample is dissolved in a suitable acid, by chlorine or by electrolysis. The carbides in the residue are broken down by oxidation and the remaining non-metallic contents are examined by the methods of micro-analysis.]

FIG. 75. Chromium inclusion (stria or 'Schliere') in steel type 100Cr6, showing effects of two impacts of infra-red laser radiation. Crater diameter 40 μm.

The methods of metallographic analysis described by Klinger and Koch[62] are well known (see also Gegus).[38] Although these methods have been specified in detail for a great number of different iron alloys they do not by any means meet all requirements, quite apart from the difficulty of the procedures and the long time they require. Hence there are many possibilities of using spectrochemical analysis with a laser source (Figs. 75 and 76) because this form of analysis is time and labour saving, does not entail loss of material and avoids possible contamination.

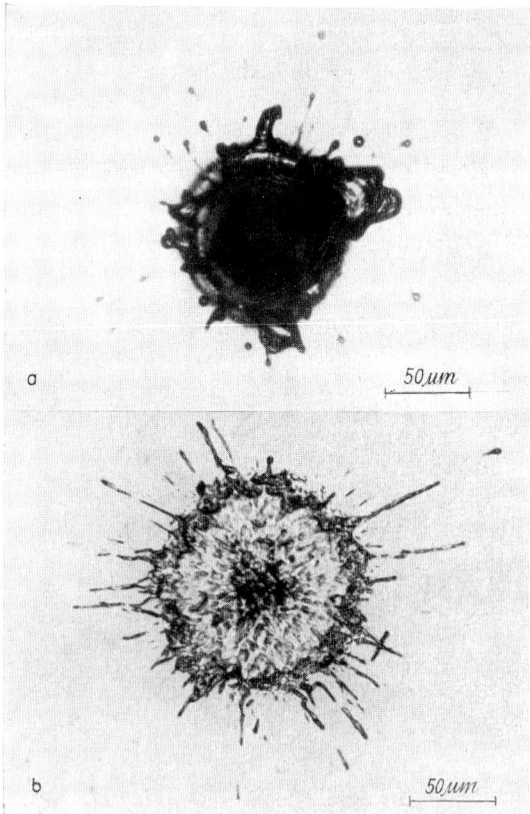


FIG. 76. Effects of laser impacts on a steel sample. Photographed with photomicrography attachment to LMA 1 model; reflected light, microscope objective ×16/0·20, projection objective ×4.

(*a*) Crater of 100 μm depth and 110 μm diameter produced by non-switched ruby resonator.

(*b*) Shallow crater of 175 μm diameter produced by Q-switched ruby resonator.

Fig. 74 shows different laser craters in a steel sample with silicate inclusions. In the spectrum resulting from a single laser firing, which produced a crater of 25 μm diameter, manganese and silicon lines were found in addition to the iron lines. The LMA 1 outfit was used with the reflecting objective ×40/0·5 and its optical adaptor with the diaphragm No. 1. A further spectrum obtained with three laser firings yielded considerably more information; the laser craters then had diameters of 60 μm. The second time the instrument was used without a diaphragm, but otherwise under the same working conditions which were:

| | |
|---|---|
| Resonator | $Nd^{3+}$ glass |
| Flash tube voltage | 2 kV |
| Capacitor bank | 508 μF |
| Electrode voltage | 1·9 kV |
| Capacitance of spark gap | 2·5 μF |
| Inductance of spark gap | 60 μH |
| Electrodes | Carbon |
| Spectrograph model Q 24: | |
| Condensing optic: $f$ = 80 mm | |
| Slit width | 15 μm |
| Aperture ratio | $f$/15 |
| Photographic plate | ORWO WU 3 |
| Developer | ORWO R 09, 5 per cent, 5 minutes, 20 °C |

The analysis showed that the following elements were present together with the iron: $Si_{ss}$, $Mn_{ss}$, $Ti_{ss}$, $Fe_{ss}$, $Ca_{ss}$, $Al_s$, $Cr_s$, $Mg_{m-s}$, $Cu_m$ (spectrum line intensity: ss = very high; s = high; m = medium).

This example indicates that with larger inclusions as much material as possible should be used for the analysis so that elements present in low concentrations can also be detected.

The two laser craters in steel illustrated in Fig. 76 were produced by a non-switched and by a switched laser, respectively; they show that the non-switched laser is preferable for the micro-analysis of metallic surfaces. The single pulses of a switched laser only act for a very short time on the sample so that the hit surface region melts and the fused particles are ejected by the developing pressure. The impact of a non-switched laser produces craters of approximately equal depth and diameter.

With the LMA 1 model one can in the non-switched mode of laser operation produce crater dimensions—diameter and depth—in the range from 10 to 250 μm. When the instrument is operated with passive Q-switching the following interesting experiment can be carried out.[88]

By employing solutions of different concentrations and therefore different transmission values up to 100 per cent one can achieve a continuous transition from single pulse working to working with a specified number of spikes up to multiple spike operation of a non-switched laser. At the same time one can also continuously increase the crater depth over a certain range because there is a linear relationship between crater depth and number of spikes within the limits from three to about twenty. The crater diameters, inner as well as outer, pass through a maximum at certain transmission values.

It is furthermore interesting to compare the craters resulting in metals at the transmission values of 95 per cent and 100 per cent, that is for a laser discharge with about twenty spikes, and laser operation without Q-switching (about a hundred spikes). At a transmission of 95 per cent the crater is of well defined, clear-cut shape and its walls look etched; but when produced by a non-switched laser the crater walls have fused edges, which are often raised, and sometimes minute fusion beads are scattered around the crater. Such features make it difficult to calculate the evaporated sample mass in quantitative analyses.

Whereas optimum substance removal for metals was obtained with a transmission of 95 per cent, the corresponding transmission values for fused silica lay in the range 78–84 per cent. This shows that Q-switching with *fixed* parameters cannot be optimized for all types of materials.

Some work on the spectrochemical analysis of steel inclusions by means of a laser microprobe has been described in literature.[60] Ryan and Cunningham investigated slags, high-efficiency refractory materials and deoxidizing agents. In the Mn–Si oxide glass of ribbon-shaped inclusions four crystalline oxides, those of Mn, Fe, Al and Ti, as well as $SiO_2$ were identified. In threadlike discontinuous weakness zones the presence of $Al_2O_3$, $TiO_2$, CaO and $SiO_2$ has been established.

McCormack combined vaporization by laser action with an etching procedure to determine oxide inclusions.

Rasberry *et al.*[105] dealt with the influence of the resonator characteristics of the Q-switched laser on the determination of Cr and Mn in low alloy standard steels and of Mn and Ni in high-temperature ferrous alloys.

Volkmann[131] discusses the case, rarely met with in practice, in which an inclusion may contain elements not occuring in the steel matrix. If elements present in the steel are also found in the inclusion the focused radiation may have penetrated a thin inclusion and may have also vaporized the steel.

## *Non-ferrous metal alloys*

*Copper and copper alloys.* The laser micro-analysis of the copper ores $CuFeS_2$ (chalcopyrite), $Cu_2S$ (chalcocite), $CuPbSbS_3$ (bournonite) and $CuCO_3Cu(OH)_2$ (malachite) (see Fig. 64) does not present any difficulties. Applied at intermediate stages in smelting and to refined copper the method allows determination of the concentration of impurities, such as Pb, Sb, Fe, Ni, Mn, Si, and others. The main alloy elements of copper are

Ni, Zn, Sn, Al, Si, Mn, Be, Ag and Au. Suitable spectral lines are given in § 4.2.

Pantelejev and Jankowski[63a] looked into the problems of the experimental conditions for examining copper alloys by laser spectrochemical analysis; according to their report the interference by other elements can be prevented.

*Brass.* There are various types of brass containing, in addition to the main components Cu and Zn, other elements, namely Pb, Ni, Mn, Fe, Sn, Al and Si; these are added in different quantities to improve certain properties, such as strength, resistance to corrosion, working properties and hardness.

The spectrograph type required for laser micro-analysis of this material is the medium quartz prism instrument like model Q 24.

*Bronzes.* Tin bronze, aluminium bronze, leaded bronze, and any other special bronzes consist of at least 78 per cent Cu with additions of Sn, Al, Pb, Ni, Mn, Fe, Si and Be.

The spatial distribution of the alloy elements within the structure depends on the quality of the bronze melt. This distribution in turn determines the suitability of the bronze required for making bearing seats, springs, tubes, gear rims, etc.

Leaded bronze is used amongst other purposes for sleeve bearings subjected to heavy wear. In these cases exceptionally high purity of the alloy and uniformity of the lead distribution are required. The presence of impurities, namely antimony, arsenic and zinc, leads to increased segregation in leaded bronzes.

Segregated components in bronze structures can easily be examined with the laser micro-spectrum analyser and from the results the type of the impurity can be deduced. For such applications too the spectrograph model Q 24 and laser output energies of about 0·2 J are adequate.

*Zinc and zinc alloys.* The most important ores for zinc production are sphalerite (ZnS) which contains amounts of Fe, Cd, Pb and Cu varying within a fairly wide range of concentrations; furthermore zincspar ($ZnCO_3$) containing also Mn, Fe, Ca, Mg, Cd and other elements, and the silicate willemite ($Zn_2SiO_4$).

The principal components of zinc alloys besides the zinc are aluminium and copper. Magnesium is added in small quantities for protection against selective corrosion due to impurities, chiefly lead, cadmium, bismuth, and tin.

As for all non-ferrous metal alloys discussed so far either neodymium glass or ruby resonator rods can be used in the solid-state laser. When, frequently, a considerable concentrations of volatile elements are determined the quartz protection plate of the microscope objective should be regularly cleaned of the metal traces deposited on it.

*Lead and lead alloys.* The most frequently occuring lead ore is galena which contains 86·6 per cent Pb and 13·4 per cent S. Silver is also present, in concentrations up to 1 per cent and sometimes more, either in isomorphous mixture as $Ag_2S$ or $AgBiS_2$, or as an additive in form of different silver ores. The contents of Zn, Fe, Cu and Sb vary. In galena coming from Freiberg Ag and Cu are easily found with a single laser firing. These impurities are still present in the crude lead during smelting. The total contents of impurities after refining are less than 0·1 per cent.

The main alloying elements for lead are tin and antimony. These two elements form cubic SbSn crystals. Babbitt metals also contain up to 5 per cent copper which appears in the form of $Cu_6Sn_5$ crystals.

*Aluminium and aluminium alloys.* The raw material for aluminium production is bauxite, a mixture of minerals in which amongst other compounds $Al_2O_3$, $Fe_2O_3$, $SiO_2$, $TiO_2$ and $V_2O_5$ are present. The binary and multicomponent alloys of aluminium contain Si, Cu, Mg, Mn, Zn, Fe, Ti, Sb, Ni and Cr.

A great variety of etching methods exists for identifying the many intermetallic compounds of aluminium. For such determinations the fast and simple method of laser micro-spectrochemical analysis can close a gap.

*Magnesium and magnesium alloys.* Magnesium occurs in nature, amongst other forms, as carbonate, fluoride and chloride. The main impurities are Fe and Al. Magnesium alloys are formed of Al (up to 10 per cent), Zn (up to 4 per cent), Si (up to 1·5 per cent), Mn (up to 2 per cent), and furthermore Ag, Bi, Cu, Zr, Cd and Ce.

*Gold–nickel alloys.* In 1966 the authors were asked to examine the distribution of the Au and Ni components across a gold–nickel welding seam. This was done with a series of laser pulses. The spectral lines Au 242·8 nm and Ni 241·6 nm were used for the analysis with photometric evaluation (Fig. 77). The reference standards required for the semiquantitative analysis of such metal samples are fairly easily produced; in preparing the standard melts, however, great care must be taken to achieve the highest possible homogeneity.

*Platinum alloys.* Also metal working industries often have to face problems which arise from relatively unimportant causes, but which can have far-reaching effects. Once two instrument parts were submitted to the authors' laboratory which presented identical macroscopic and microscopic appearances, but completely different electrical behaviour. The components had platinum contacts one of which on examination with the LMA 1 laser micro-analyser was found to have a high nickel content. The origin of nickel contents in platinum is easy to understand because there are on earth several platinum deposits where, in close paragenesis, nickel ores also exist.

This example of a successful application of laser micro-spectrochemical analysis shows that a routine test of platinum components is always advisable. In the above case, the contact units could be examined without dismantling and without impairing the subsequent performance of the part without nickel contents.

*Zircon–uranium compound alloys.* Cerrai and Trucco[24] combined metallographic methods and laser micro-spectrochemical analysis for

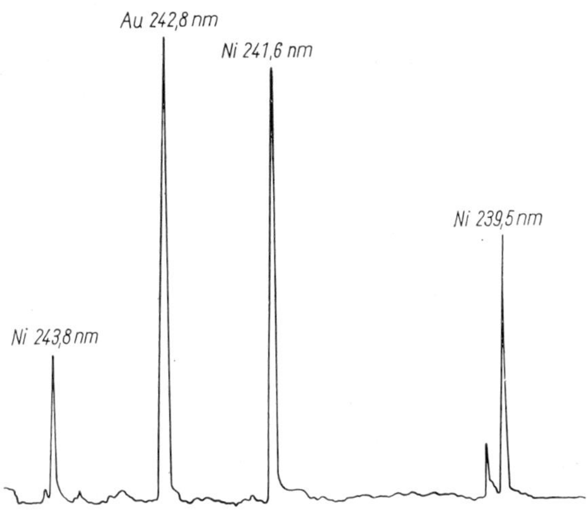


FIG. 77. Microphotometer trace of section of spectrum of Au-Ni welding seam. Instrumentation: outfits LMA 1 and Q 24. Crater diameter 60 μm.

establishing tungsten micro-diffusion in uranium oxide rods protected by Zircaloy 2. Similarly, the laboratories of CISE, Segrate, Milan, were able in the course of testing nuclear fuels to determine by laser irradiation the Fe, Cr, Ni and Mn contents of minute steel inclusions in sintered uranium oxide.

### *Smelting preparation products*

As already mentioned, a laser spectroscopic analysis can be carried out on all raw materials and intermediate and end products of metal production. The examination of ores, in particular with consideration of the following preparation for smelting, has been discussed in § 4.1.1.

In carrying through the processes of preparation regard must be paid to the degree of breakdown, enrichment or depletion. For this purpose a method of rapid analysis as routine production control is highly desirable. A comparison of grain sizes and the diameters of the areas of impact of focused laser radiation will show the possibility of employing laser micro-spectrochemical analysis for examining crushing, grading and sorting fractions: the diameters obtained with the Zeiss model LMA 1 fitted with the reflecting objective ×40/0·50 or the achromat ×16/0·2 can vary between 10 μm and about 250 μm. The sifting fractions of a sulphide ore being prepared for smelting have the following grain size distribution:

| | |
|---|---|
| 45–55 μm | 17·6 per cent |
| <85 μm | 76 per cent |
| 125< μm | 91 per cent |
| 200–240 μm | 0·4 per cent |

[*From Handbuch der Mikroskopie in der Technik (Handbook of Microscopy applied to Technology)*, Vol. II, Part 2, p. 416.]

*Smelting slags*

The economic significance of smelting slags is evident from the fact that in producing 1000 tons of raw iron in a Thomas furnace 600 tons of furnace slags are formed. Metal furnace slags are solidified fused silicates. The method of examining them with a laser micro-spectrum analyser is therefore the same as that for rock samples, described in § 4.1.1.

The chemical nature of the slags follows from the types of ores used, the additives and the course of the smelting process. Apart from the overall chemical composition, the proportion of non-ferrous metals and the distribution of the elements detected in the analysis are of main interest. Slags can be tested in any form, without preparation or on polished sections, or in the form of polished thin slices. In the latter case the chemical analysis can be carried out together with the mineralogical determination which will practically exclude any possibility of mistakes in the findings.

In most cases it is not possible to determine silicates completely in polished sections. It has previously been recommended to prepare a polished section of the slags for analysing the opaque minerals; but this is by a no means simple. Ramdohr advises not to apply etching methods uncritically. Because of the closely knit intergrowth microchemical methods are only occasionally successful. Microscopic examination by itself fails for purely vitreous or fine-crystalline slag formations. Such samples are best systematically examined with the LMA outfit and with the spectra of the vaporized small sample quantities recorded one below the other on the photographic plate. The vernier scales of the simple mechanical or of the coordinate stage of the LMA microscope allow exact control of the sample movement and an exhaustive survey of the material. The slag constituents to be found are pure elements, sulphides, arsenides, antimonides, oxides, silicates, phosphates and vitreous substances.

The following elements can be identified in slags by micro-spectrochemical analysis: Fe, Cu, Pb, Ag, Zn, Mn, Mg, Ca, Si, Al, Cd, Sn, Sb, Ti, Cr, Co, Ba and others.

### 4.1.3. Chemistry and silicate technique

An economically very important field of application of the laser micro-spectrum analyser was brought to the authors' attention by the letter from a customer who wrote:

'Die Funktionstüchtigkeit und Lebensdauer hydraulischer Geräte ist von der Sauberkeit der Arbeitsflüssigkeit (Druckübertragungsmittel) abhängig. Die Verschmutzung der Flüssigkeit wird u.a. bei uns mikroskopisch ausgewertet, d.h., es wird eine Flüssigkeitsprobe durch Membranfilter bzw. Laborfilterpapiere filtriert, und der auf der Filteroberfläche abgesetzte Rückstand von Feststoffteilchen wird mikroskopisch nach Teilchengröße und -zahl ausgewertet. Auf Grund der bekannten Verschmutzungsquellen sind je nach Einsatzfall der Hydraulikanlage Industrie- and Ackerstäube, Metallabrieb, Bearbeitungsspäne, Schlacken, Rost und Verharzungen der Hydrauliköle u.a. im Filtratrückstand zu vermuten. Bei unseren Untersuchungen der Teilchen konnte eine stoffliche Bestimmung bzw. qualitative Analyse nicht vorgenommen werden. Zur Beurteilung der Ölalterung, Bestimmung der Gleitpaarung mit erhöhtem Abrieb, Auffindung von Verschmutzungsquellen müßten die Teilchen im Filtrat auch qualitativ analysiert werden...'

[The performance and lifetime of hydraulic instruments depend on the purity of the working fluid (pressure transmitting medium). Apart from other measures we estimate the contamination microscopically; a sample of the fluid is passed through a membrane filter or a laboratory filter paper and the residue of solid particles lying on the filter surface is examined by microscope for particle size and number. Possible sources of contamination have been established by experience and accordingly any of the following substances appropriate to the particular application of the hydraulic system may be expected to be present in the filter residue: industrial and agricultural dust, metal wear products, shavings and turnings of processed materials, slags, rust, resinifications of the hydraulic oils and so on. We could not in our tests on the particles identify the substances or carry out qualitative analysis. To judge deterioration of the oil through aging, to trace pairs of sliding members with excessive wear, and to spot sources of contamination, one would have to analyse the filtrate particles also qualitatively.]

The authors have carried out a number of such spectrochemical analyses and found it advisable to take photomicrographs before and after the analyses so that the results could also later be interpreted as accurately as possible. For this work the LMA 1 model was used with a quartz spectrograph Q 24. The laser was fitted with a neodymium glass rod and the objectives were the types $\times 40/0{\cdot}50$ and $\times 16/0{\cdot}20$. The electrical operating conditions were: flash tube voltage 2 kV; capacitor bank 508 $\mu$F; spark gap–voltages of 0·6, 1·0 and 1·3 kV, capacitance 2 $\mu$F, inductance 60 $\mu$H. The spectrograph worked at an aperture ratio of $f/15$

and with a slit width and length of 15 μm and 2 mm respectively. The discharge was directly imaged on the slit with the condenser of 80 mm focal length. No difficulties were encountered in these analyses. The detected elements and their quantities varied greatly; in the main Ca, Mg, Zn, Al, Cu, Si and Fe were found.

Polyester film is the carrier material of magnetic tapes for sound recording, of the video tapes for magnetic film recording, of the computer tapes for electronic data processing equipment, of the instrument tapes for analogue measurement recordings, etc. If the film is to be suitable for these applications it must be perfectly uniform in thickness and be free from any impurities, because any irregularity would spoil the tape. Orders received by the authors indicate that these requirements pose serious problems for the producers; it is either difficult to keep the plants completely dust free, or the catalysts used disturb the processes. To the authors' knowledge inclusions in the film could not at that time be analysed by any other method, such as X-ray fluorescence spectroscopy or electron beam micro-analysis, because the thin non-conducting samples are very difficult to prepare for such tests.

The specimens submitted to the authors were examined under the microscope of the laser micro-spectrum analyser with and without polarizers. A variety of inclusions was found; an example is shown in Fig. 78. By far the greatest number contained calcium, probably present in form of an organic compound. In others iron, magnesium and silicon were detected. As parts of the film were only 6 μm thick, and as in some cases the inclusions were linked in a chain-like formation, three to four laser bursts were used for a spectrum. The instrumentation and operating details were:

Laser micro-spectrum analyser model LMA 1 with $Nd^{3+}$ glass resonator rod

| | |
|---|---|
| Flash tube voltage | 2 kV |
| Capacitor bank | 508 μF |
| Spark gap: Voltage | 1·5 kV |
| Capacitance | 2 μF |
| Inductance | 60 μH |
| Reflecting Objective | ×40/0·50 |
| Quartz spectrograph model Q 24: | |
| Condenser | $f = 80$ mm |
| Slit width | 14 μm |
| Slit length | 2 mm |
| Aperture ratio | $f/15$ |

In a nuclear physics laboratory a neoprene sleeve was used as a component of a bubble chamber. Small glass and quartz fragments which were embedded in the surface of the sleeve and impaired the performance of a sliding mechanism made of metal could be detected by laser micro-spectrochemical analysis besides the zinc filler contained in the plastic material.[24]

The use of micro-spectrochemical analysis for examining natural crystals has been dealt with in § 4.1.1. As the result of scientific and technological developments many crystal types can now be artifically grown; the purpose of these processes is to produce materials as pure and homogeneous as possible. The chemical composition and the distribution of the

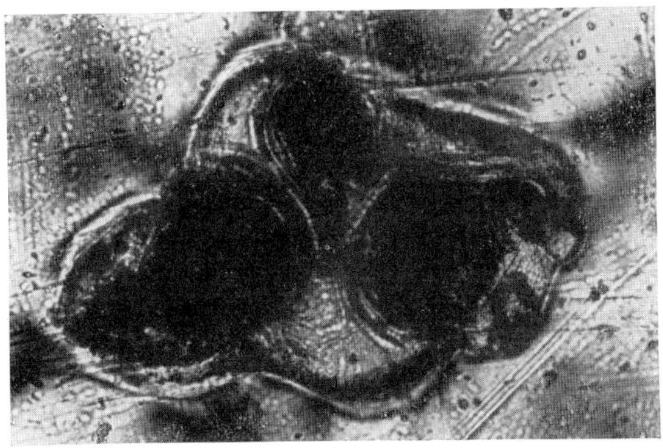

FIG. 78. Micro metal inclusions in polyester foil vaporized by focused laser radiation.

constituent substances can be systematically controlled by routine analysis of microgramme sample quantities. Any such analysis only takes fractions of a second and the results are permanently recorded on a spectrum plate, conveniently available for comparison.

Impurities and inclusions can immediately be analysed in a spectrum taken for preliminary inspection and thus it becomes possible to establish the cause of a defect. Often changes of the physical properties of solids are due to variations in the chemical composition in microzones.

Vaporization by a non-switched laser is satisfactory for the analysis of the inclusions themselves and also of many coloured and opaque glasses or crystalline matrices. A Q-switched laser, however, is necessary for analysing glasses transparent for the laser wavelength and especially for comparison with the composition of inclusions. As described in § 3.2 the LMA 1 outfit can, with a cell attachment for a saturable absorber, also be operated in the Q-switched mode. The authors examined various glass types of the VEB Schott and Associates, Jena, with regard to removal of material by laser irradiation.*[88] Various filter glasses were selected according to their transmissivities at the wavelengths 700 and 1050 nm and

* The experimental work was carried out by the authors' co-worker, Dipl.-Phys. J. Mohr.

arranged in groups of the same types. In the group BG 5, 9, 14, 27, 3, 29, 33, in this order, the transmissivity for the neodymium glass laser radiation increases from 0·13 to 0·95; and in the group BG 28, 23, 14, 16, 1, 22, 21, 24, 20 for the ruby laser radiation from 0·02 to 0·95. The glasses of these groups, in the same order, were also irradiated with the two lasers in non-switched operation, and in both cases the crater diameters were found to be clearly proportional to the absorption coefficients of the glasses. Glasses were then selected with transmissivity 1·00 for the first wavelength (700 nm), namely types GG 3, OG 3, RG 1, WG 1, and for the second wavelength (1050 nm), namely GG 5, OG 5, RG 5, WG 5, and compared with glasses of a transmissivity of less than 0·03 for the same two wavelengths, namely types UG 1, VG 12, BG 28, RG 7, and types BG 30, VG 3, NG 9 respectively. All these investigations showed unmistakably that transparent materials can be vaporized with a non-switched laser only if their transmissivity for the laser wavelength is very much less than 1·00. With the Q-switched ruby laser, on the other hand, the form of material removal in all examined filter types varied only little in crater diameter and depth compared with the results for the non-switched laser.

Fused silica proved to be most resistant to focused and unfocused laser radiation; for this reason it was chosen as the material for the total internal reflection prism for the Q-switch mode operation of the ruby laser resonator unit. Different from all tested glass types, fused silica has a definite maximum of material removal in the transmission range of the Q-switch liquid from 78 to 84 per cent.

Unlike flint glass, crown glass cannot be vaporized with non-switched laser radiation at all or only with great difficulty. In experiments undertaken with the laser model ZFL 1000 (output energy 1 J) on thirteen glass types (borosilicate, fluor, phosphate, barium, zinc, dense, extra dense and ordinary crown glass) with well polished surfaces no vaporization of material occurred. The extra dense type SSK 6 could be vaporized with a more powerful neodymium glass laser of more than 1·5 J output energy; craters of 100 to 200 μm diameters and depths of half these values resulted. When the surfaces were roughened through scratching or etching the ordinary and fluor crowns K 6 and FK 4 could also be vaporized, but only small shallow craters of about 40 μm diameter or less were produced; type FK 4 in fact only showed circular burn marks of about 50 μm diameter after irradiation with the non-switched laser. But material could be removed without exception from all crown glasses by using a ruby laser with passive Q-switch. The crater dimensions varied between the glass types within the following values: external diameter 20 to 100 μm, internal diameter 5 to 40 μm, depth 0 to 10 μm.

Similar experiments on substance removal carried out with the laser model ZFL 1000 on seven different flint glasses (crown, barium, dense, dense barium, light barium, extra light and short flint) gave more positive results. Only in the case of some of these flint glasses (barium, light barium, extra light and short flint – code numbers BaF 10, BaLF 5, LLF 6 and KzF 4) could no material be removed from a clean polished surface; yet other samples of the same types (namely BaF 3, BaLF 4, LLF 8, KzF 2) showed vaporization.

It again has to be stressed that laser micro-analysis is a spot measurement method intended for examination of inclusions and similar small regions. This special technique of emission spectroscopy is not called for in normal routine analyses of homogeneous glass.

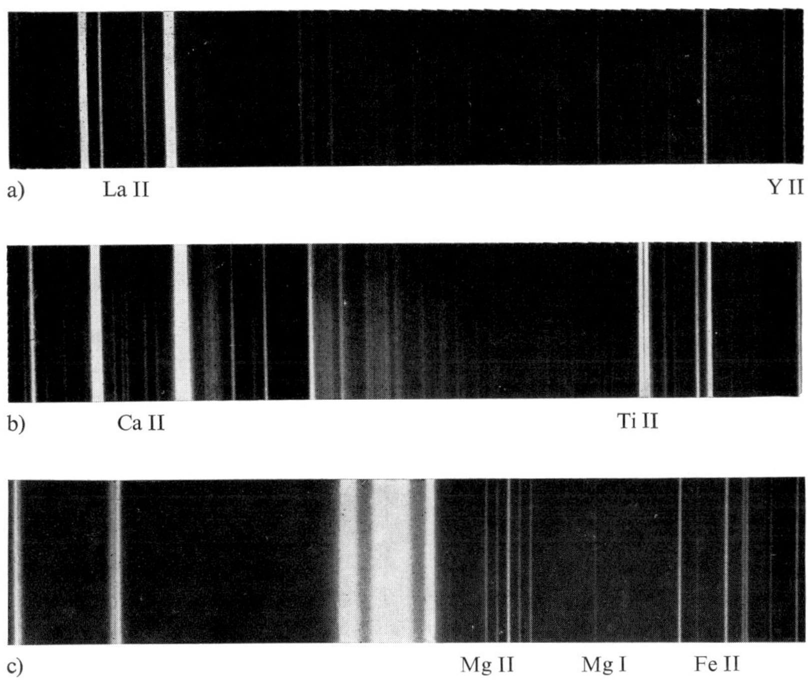


FIG. 79. Silicate analysis.

(*a*) Identification of Y and La in orthite, in the presence of Ca and Al.
Analytical lines: Ca II 393·37, 396·85 nm; Al I 308·22, 309·27 nm; La II 394·9 nm; Y II 371·03 nm.

(*b*) Determination of titanite cations Ca, Ti and La in long wavelength section of working range of spectrograph Q 24.
Analytical lines: Ca II 393·37, 396·85 nm; Al I 308·22, 309·27 nm; La II 394·9 nm; Ti II 368·52, 375·93, 376·13 nm.

(*c*) Identification of olivine cations.
Analytical lines: Mg II 279·08, 279·55 R, 279·81, 280·27 R nm; Fe II 273·96, 275·01, 275·75 nm; in between the characteristic group of Mg I lines around the strongest component 277·98 nm. (R = self-reversed.)

But if in certain cases, as mentioned before, an inclusion has to be compared with the surrounding matrix and if no Q-switched laser is available, an expedient can often be found, as for example roughening or etching the surface, coating it or smoking it by discharging the electrodes close to the surface without firing the laser.

Special reference should still be made to the important group of silicates used in technology. Laser micro-spectrochemical analysis here offers considerable advantages compared with the methods, available so far, of X-ray technique, microscopy, electron microscopy, infra-red spectroscopy, conventional wet chemical analysis and spectrochemical analysis on the macroscale. These methods are not to be replaced, but considerably supplemented.

Laser micro-spectroscopy can be applied by the glass manufacturing and glass working industries in the examination of certain features, briefly discussed in the following, for the purpose of finding the causes of technical manufacturing faults.

*Solid inclusions*

1 First examination immediately on vitrification of the mixture being fused, especially in the case of trial melts. Of interest here are the still unmolten ingredients (melt residues) which convey information on the course of the melting process and allow conclusions to be drawn on the behaviour of the raw materials.
2 Devitrification products (so called stones, knots and crystalline segregates).
3 Foreign bodies such as non-molten particles of the material of the melt containers and tools (tanks, pots, floaters, bridges, etc.)

The method is not applied to the examination of gas bubbles unless they contain crystallized material. It also has to be noted that, as in the only method of spectrochemical analysis previously practised, the cations present in the structure are easier to detect, which can be seen from Table 16. In the special case of segregation of sulphates this means that sulphur and oxygen cannot be determined.

The ultimate purpose of investigations into glass defects is to find the cause of the faults. Besides the mentioned usual routine methods and electron beam micro-analysis,[29, 67, 134] laser micro-spectrochemical analysis can now be adopted as a procedure combining microscopy and emission spectroscopy in the microgramme range. Ryan, Ruh and Clark[114] were the first to examine glass defects with a laser microprobe which they did quantitatively on an $Al_2O_3$ knot, on a stone with high $Cr_2O_3$ concentration (35 per cent $Cr_2O_3$, 30 per cent $Al_2O_3$, 21 per cent $Fe_2O_3$, 12 per cent MgO) and its surround and on a zircon nepheline inclusion.

Similarly $SiO_2$ stones, barium silicates, $BaO–B_2O_3$ silicates, $ThO_2$ crystals and similar samples can be analysed. A complete mineral determination is therefore possible by combining laser micro-analysis with a second procedure, for instance infra-red absorption spectroscopy.[16, 79]

Fig. 80 shows a sample of special glass with inclusions of spherical beads. Analysis with the LMA 1 outfit revealed zinc and silicon in high concentrations as the main constituents besides a smaller calcium content.

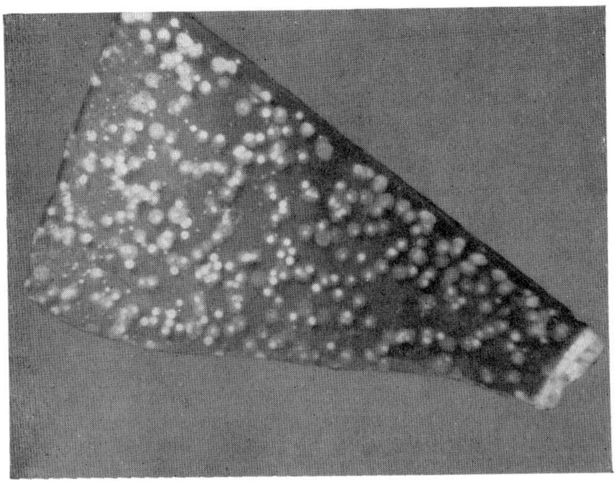

FIG. 80. Typical analytical problem for laser micro-spectrum analyser: special glass type with bead inclusions.

Laser micro-spectroscopy can be the means for finding the causes of manufacturing failures, for instance in the production of fused silica melts. The examinations of two melts with the LMA 1 analyser gave the following results:

| *Melt with metal beads* | | *Melt with foreign glass matter* | |
|---|---|---|---|
| Fe | ss | Si | ss |
| Si | ss | B | ss |
| Ca | s | Ca | s |
| Cu | s | Cu | m |
| Mg | s | Mg | m |
| Al | m | Ba | m |
| Cr | m | Al | m |
| Mn | w | | |
| B | w | | |

ss: Very strong line intensity
s: Strong line intensity
m: Medium line intensity
w: Weak line intensity

The small metal beads in one of the melts (Fig. 81) consisted mainly of iron, with some chromium and manganese content, which was introduced into the fusion mixture through abrasion in the breaking down of the raw material in a jaw crusher. At the fusion temperatures the iron scrapings melt into small beads or, if only present in small quantities, cause black spots. Ca, Mg, Al and B come from the crusher previously used for other purposes and not cleaned thoroughly, and the Cu content comes from the bronze sieve used at one stage of the repeated grading of the crushed material. Silicon is a component of the fused silica.

The foreign matter of the second melt was duran glass which inadvertently got into the raw material. Duran glass contains all the detected elements B, Ca, Mg, Al and Ba. Silicon again is a constituent of the fused silica.

The intense blackness of the inclusions is probably caused by $As_2O_3$, which is present in the duran glass and, in the strongly reducing atmosphere, is reduced to As. No As lines could be found in the spectrum because the concentration is below the detection limit.

The authors have carried out another rapid analysis with the LMA 1 outfit on a very different sample from the glass industry. In the fragment

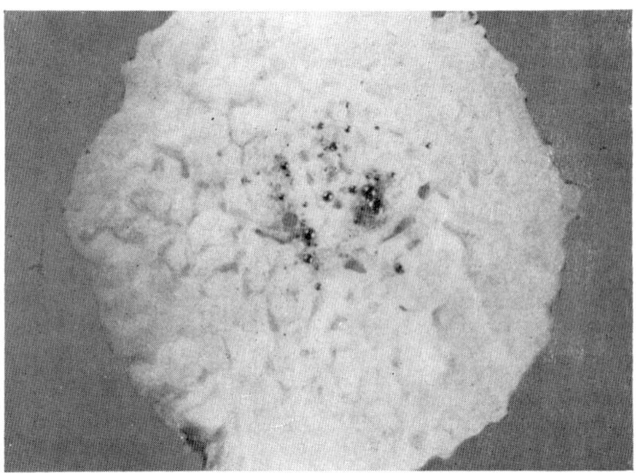

FIG. 81. Vitreous silica melt with small metal beads which can be examined with LMA 1 model without sample preparation.

from a melting pot (Fig. 82) containing 1·2 per cent $Fe_2O_3$ and 1·4 per cent $TiO_2$ black and white dots of various sizes could be seen embedded in the matrix. The problem was to examine the composition of these inclusions compared with that of the matrix. The laser micro-spectrum analyser fitted with a flat-field achromat $\times 16/0{\cdot}2$ was used in conjunction with a quartz spectrograph model Q 24, with carbon electrodes and the following settings of the supply units: flash tube voltage 2 kV, capacitance bank for laser 381 μF, electrode voltage 1·5 kV, capacitance of spark gap 2 μF, inductance of spark gap 60 μH.

The three spectra of the matrix, the dark pigments and the white dots all showed equally low iron contents, but those of the white dots a higher concentration of silicon and aluminium. The calcium concentration in the

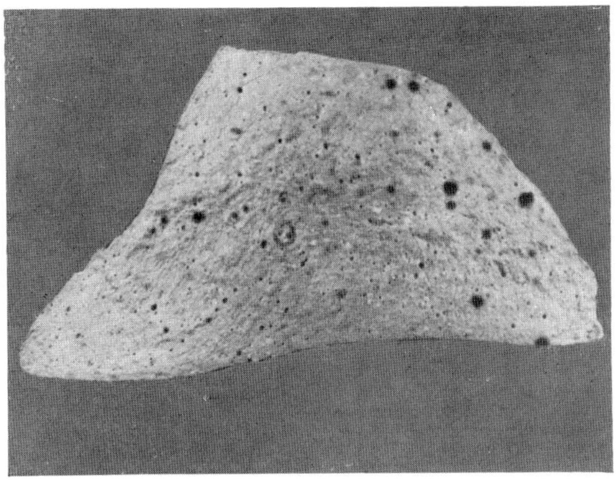

FIG. 82. Fragment of melting pot with light and dark inclusions.

white dots is lower than in the matrix, but higher than in the black inclusions. The latter also show besides the high calcium content a higher concentration of zirconium and chromium. The quantity of titanium is greatest in the matrix and of equal lower value in both types of inclusions.[83]

In the glass working industries problems often arise in connection with glass to metal joints, as for instance with valves and lamps or connection of electrodes through a vacuum seal. Such samples of unusual shapes can be mounted on the microscope stage of the LMA 1 model and be directly analysed.

An enamelled object consists of two basically different materials. The base is normally a metal, for instance sheet steel, cast iron, aluminium, copper, tombac (red brass), nickel silver, silver, gold; occasionally also glass. The enamel coating, usually made up of two layers (ground coat and cover coat), is produced by melting and sintering and should preferably be a vitreous substance, fundamentally an oxide. Although enamel jewellery was known long before the Christian era enamel research only began a few decades ago. Up to now it was essentially restricted to wet chemical analysis, microscopy, gloss and whiteness measurements, surface quality and expansion tests. Here are further promising possibilities of the application of analytical micro-methods:

1 Enamel raw materials. Testing and control were previously carried out by conventional chemical analysis. The purpose of the examination is to find impurities in quartz, felspar, fluorite, calcite and dolomite (the carbonates are identified by staining), clays, borax, sodium carbonate, potassium nitrate. Of special importance are colorants, glazes, and opacifiers such as Pb, Ba, Ni and Co oxides, $SnO_2$, $TiO_2$, $ZrO_2$, $ZrSiO_4$, $CeO_2$, ZnS, Pb–As silicate, $Sb_2O_3$, $Ca_3(PO_4)_2$, NaF, $CaF_2$.

With laser micro-spectrochemical analysis the process of raw material melting and the various stages of enamel production can be controlled. Of interest is the distribution of the chemical elements in the melt, granules and in the contact surface between enamel and metal, and the homogeneity obtained after the addition of bonding oxides, flux, colorants and opacifiers.

2 As in glass production, here too the introduction of contaminations from the materials of the muffles (fire clay, corundum, silicon carbide) and the furnace lining (fire and ramming clay) is an important factor. The analysis of dark inclusions in the ramming lining of a furnace for instance revealed the presence of ferrous compounds which cause heavy corrosion and also spoil white enamel.

3 Testing of the metal base. In the past the suitability of a material for enamelling was judged by the results of chemical analysis and of microscopic examination of the surface, undertaken separately. These two procedures can now for the first time be combined by means of laser micro-spectroscopy: a spectrochemical analysis can immediately be made of surface parts of relatively large samples which appeared to be of interest in observation under the microscope. Previously the inclusions were isolated as described by Klinger and Koch.[62] The following tests according to Dietzel[35] consisted of microscopy for determining shape, colour and optical properties, hardness measurements and chemical analysis.

Apart from checking the metal base before enamelling, it often also becomes necessary later to search for the cause of faults which have developed because blemishes in the sheet metal usually show up on the surface of the enamel. Possible components of slag inclusions in sheet steel are: Fe, Mn, Si, Ca, Ni, Mo, Cr, Al, Be, Ti and Zr. A copper content in sheet steel of more than 0·2 per cent is known to be harmful because it aggregates on the sheet surface during pickling and renders the bonding oxides of the ground coat enamel ineffective.

Another new possibility is the application of the laser micro-method in the ceramics and cement industries for controlling the raw materials, the works processes and the finished products. Coarse ceramics (building ceramics, coarse earthenware, silicate and non-silicate refractory ceramics) and fine ceramics (pottery, technical ceramics, porcelain and ceramic coverings) can be analysed as well as the mineral contents of cement clinker.

The raw materials for the products of the refractory industries are natural clays and bauxites; their impurities collect during the melting in certain regions and are liable to cause faults when the material is processed in melting tanks and furnaces. At the high temperatures reactions often take place between the furnace lining and the contents.

During the distillation of zinc willemite, zincite and spinels can be formed in the fire clay lining. In the vault of a lead refinery furnace various lead components may be found besides gahnite and willemite, also Ti, Fe, Mn and Cu. Besides the main constituents Si, Al, Mg, Ca, Cr, Zr and B, the minor components and sometimes also the trace elements are of significance especially in refractory materials.

Alongside chemical analysis microscopic examination also has to a certain extent already been established as a means for quality control. It has, however, only now become possible to combine the two methods in

FIG. 83. Bakor 33 refractory sample. Laser micro-analysis showed Fe, Mn and Ca enrichments in the minute crystal inclusions.

one operation and particularly to determine by micro-spectrochemical analysis with a solid-state laser the segregations, zones, veins and inclusions visible under the microscope.

Once a refractory sample 'Bakor 33' (Fig. 83) was submitted to the authors. Its chemical composition was given as[37] 50 per cent $Al_2O_3$, 15 per cent $SiO_2$, 27–32 per cent $ZrO_2$, 1·6 per cent alkalies, 2 per cent alkaline earths, with additional small Fe and Ti contents; the mineralogical composition as baddeleyite, corundum and glass. Brown dots in crystals of flesh-like red colour were to be examined by laser micro-spectrochemical analysis. The elements Al, Si, Zr, Mg, Ca, Fe, Ti and Mn were identified in the spectra. The brown dots contained enrichments of Fe, Mn and Ca, whereas their aluminium concentration was lower than that of the flesh coloured crystals. No difference could be found in the Si, Mg, Ti and Zr contents.

The instrument of Zeiss, Jena, is also suitable for the examination of fine ceramics.[83] In discolorations, segregations, blemishes and vitrifications in porcelain, pottery or objects made of mullite, corundum, talc, titanium and magnesium oxide, cations can be found in a concentration different from that in the matrix.

A steatite article made of talc with a BeO addition has increased mechanical strength and shows the formation of needle-like crystals which, it was assumed, might be bromellite. The beryllium in these crystals of 15 to 20 μm could without difficulty be determined by the routine method of laser micro-spectroscopy because this type of analysis is also practicable for the very light elements with low atomic numbers.

Work of a new type is the examination of bonding agents by laser micro-analysis. In the production of cements from calcite, clay or bauxite the possibility of contamination by iron and other elements cannot be excluded. If oxides such as $Fe_2O_3$, $TiO_2$, BaO, $Cr_2O_3$, $V_2O_5$, $B_2O_3$, $Mn_2O_3$ or others are present in Portland and alumina cements besides the calcium aluminates, they affect the sintering as well as the setting process. Up to now the phase of Portland cement clinkers was investigated mainly by petrographic, X-ray and chemical methods. For the first named the samples are prepared in the form of thin polished slices, pieces with a polished section or powder. The microscopic tests with natural and polarized light can also be carried out with the microscope model LM of VEB Carl Zeiss, Jena. Structure components that are of interest and previously could not easily be identified within the sample can now be analysed spectrochemically without any serious interruption of the microscopic work.

Another promising field of application of laser micro-analysis may be the brick industry. Here wrong additives or unsuitable raw material composition may reduce the strength or weather resistance of the bricks and cause considerable financial losses. The authors analysed some brick samples with the LMA 1 laser instrument; white inclusions contained Si as was assumed, probably in a quartz bond. Greyish-black fracture surfaces showed very high Si and Mg content and a high concentration of Al, Ca, Fe and Ti. The Cr spectrum line 357·9 nm was of medium intensity, and Mn gave low intensity lines. The samples were not prepared in any way. The colouring was assumed to be due to the four last named elements present in some form. To establish the nature of the compounds it was suggested to supplement the laser micro-spectrochemical analysis with infra-red absorption spectroscopy and X-ray diffraction tests.

The new high resistance NUCON bricks of the Harbinson–Walker Refractories Co., Pittsburgh, Pa, USA, were already routine-tested at the research and development stage with modern analytical instruments including the laser probe.[54]

The following detection limits in terms of gramme per ton were reported for an $Al_2O_3$ powder matrix analysed with vaporization by a Q-switched laser and with a 40 J spark excitation:[132] Be 0·8; Sr 10; B 11; Pb 30; Sn 50; Ti 20; Cr 11; Fe 45; Co 24; Ni 25; Mo 21; W 400 g/t. The sensitivity is generally improved for spectra obtained with two or three laser pulses; some of the limits are then: Be 0·4; B 0·8; Ni 2; Fe 3, Cr 3, Sr 5, Ti 5 g/t.

### 4.1.4. Archaeology and history of art

That spectrochemical analysis is considerably more suitable than conventional chemical analysis for the examination of art relics has been established in numerous publications (Pittioni).[100] The great speed of the analysis, the small quantity of material required for the test and the practically destruction-free operation are the most important features of emission spectrochemical analysis that helped to increase our knowledge of the history of civilization.

The first spectroscopic investigations in the field of prehistoric studies were those of Witter and his Hallenser study group in 1931. From 1933 to 1935 spectrochemical research was also carried out and described in the USSR by W. W. Danilévskij in Leningrad. In the next thirty years such work was undertaken, partly on a large scale, by amongst others Pittioni and his circle in Vienna, furthermore in Stuttgart, Stockholm, Helsinki, Prague and Budapest, also in Poland, Italy, France, Belgium and England.

Through use of the spectrograph the conventional methods of metallographic tests and mineralogical research were, especially in this field, continuously refined with the aim of reducing the required sample quantity. Quite different procedures for spot analysis were developed, some necessitating tedious sample preparation and others requiring expensive instrumentation. Examples of these procedures are:

1 Restricting the spark spot through reducing the distance of the electrodes from the sample surface. Exact localizing is not possible here.
2 Masking the sample with insulating material, for instance a small perforated quartz or mica plate or varnish (Geyer, Doerffel and Kirst[43]). With insulating varnish, crater diameters of 0·25 mm are obtained. Two neighbouring spark spots must be 0·75 mm apart.
3 Thread-like spark discharge according to J. J. Martin. A special set-up with quartz capillaries makes it possible to reduce the area hit by the spark to dimensions of the order of 0·03 to 0·05 mm.

4 About 0·2 g of material is removed from metal samples with a small drill. The drillings are melted down to a button from which two small electrodes are formed, then filed to shape and rounded off. As prehistoric samples often are completely oxidized such sample preparation is not always feasible. As an alternative the drillings can then be filled into carbon electrodes.

A number of modifications of these methods are known.

Although highly developed and often described in literature, these methods could not on account of their disadvantages, such as necessary elaborate special sample preparation and cost of instrumentation, displace conventional chemical analysis. As can be seen from Fig. 84, methods requiring much sample material are still applied in modern work.

Micro-spectrochemical analysis with laser excitation is a significant advance. The sample need not be prepared at all. The area to be analysed can be selected and examined under the microscope. The impact of the focused laser radiation forms a crater of about 50 μm diameter, hardly visible with the naked eye, and vaporizes a mass of about $10^{-6}$ g of sample material. This quantity is sufficient for a complete preliminary

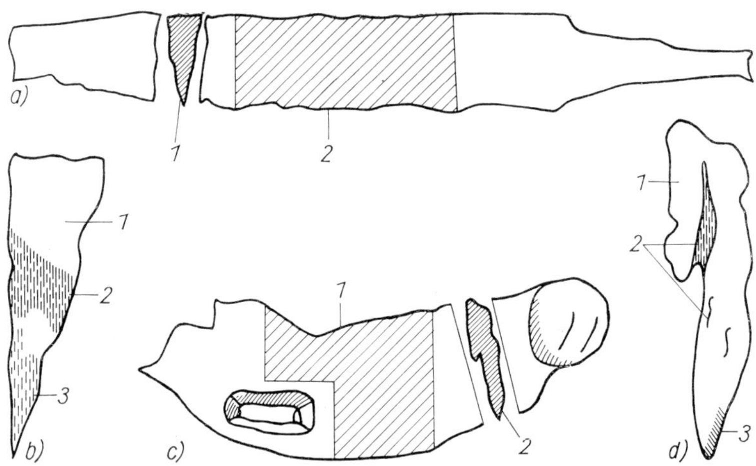


FIG. 84. Positions of analysis targets in iron finds from Berlin-Köpenick (according to Lüdemann, Schirmer and Ebert[69b])

(*a*) Knife, about full size. 1 Polished section; 2 Target region.
(*b*) Cross-section of knife. 1 Ferrite; 2 Ferrite with large slag inclusions; 3 Tempered martensite.
(*c*) Horseshoe, about full size. 1 Target region; 2 Polished section.
(*d*) Cross-section of horseshoe. 1 Ferrite; 2 Slags; 3 Widmannstätten ferrite.

analysis in which the presence or absence of sixty different chemical elements can be established. For quantitative analysis reference standards o suitable composition can be used when available (see also § 4.3.).

As in the generally practised macro-spectrochemical analysis internal standards may also be used here. The ultra-violet spectrum of the copper in bronzes is recommended for determining Sn, Pb, Ag, Ni, Co, As, Sb, Bi (Otto).[95]

A valuable feature of spectrochemical analysis is the fact that the photographic plate forms a permanent record for later comparisons and interpretations.

The electrodes used with lasers in micro-spectrochemical analysis are independent of the sample material also in the special applications in archaeology and art history; they may for instance be carbon or some other material. The sample substance therefore may be electrically non-conducting, for instance an oxide, silicate, carbonate, etc.

For testing metal objects the same considerations hold that have been discussed before in the section on metallography. Besides the common alloys of today many others of unusual composition, however, occur in prehistoric metal relics. In these cases spectrochemical analysis has been found to be particularly helpful for detecting more or less unexpected elements in spectra taken for preliminary inspection. Knowledge gained from the history of metal techniques is of great interest for modern metallurgy. It is often possible to deduce from the element combination the place of occurrence of the raw materials. This, however, presupposes spectroscopic analysis of the materials from all deposits in question.

As a result of analytical examinations of metal finds (Otto)[95] the bronze age could be subdivided into a number of periods. The oldest copper relics consist of almost pure copper. This came from deposits in which native copper or pure carbonate or oxide ores (malachite, azurite, cuprite) occurred. Such pure copper objects rarely contain so-called key elements which allow attribution to one particular deposit. From the same time period (late Stone Age, about 2200 B.C.) also come objects made of less pure metal; today such material would be classified as crude copper, that is copper smelted from ores that were contaminated by other minerals. In these cases it may be quite feasible to establish the origin.

The earliest bronze-like relics found in Central Europe are copper arsenic and copper tin alloys. Tin bearing copper ores are known to come for instance from the Vogtland district and the Harz mountains (Saxony, G.D.R.). The tin and arsenic contents have been increased through the addition of tin ores and arsenic compounds. The use – amongst others – of copper sulphide ores (chalcopyrite, bornite) is notable because extraction of copper from them is still considered difficult even today.

Copper alloys with high Ag, Ni, As and Sb contents are nowadays practically unknown in metallurgy. The origins of such alloys are thought to be tetrahedrites used in smelting. These groups of materials can be especially well traced back to certain deposits, as for instance the district of Kamsdorf near Saalfeld, Thuringia, G.D.R. The seams of this district have barite gangues and therefore barium is a specific key element for the deposit.

As conventional spectrochemical analysis has already well proved its value in archaeology there is still much further scope for the micro-methods, not only in work on alloys of copper, lead and iron, but also on precious metals for which the advantage of small sample quantity requirements is clearly obvious. The same consideration applies to coins and similar small objects that, being unique, must only be tested by destruction-free methods.

In 1966 a report was published on the composition of very durable parts of a small 5000 years old ox from Anatolia belonging to the Boston Museum of Fine Arts.[53] High arsenic and silver concentrations were established with the laser instrument of Jarrell-Ash, producing in the sample surface craters of about 200 μm diameter. It is of greatest interest to see that the museum administration consented to the semi-micro-spectroscopic examination of an object worth $ 40000.

As a laser micro-spectrochemical analysis can equally well be carried out on non-conducting samples the examination of glasses, pearls, glazes, paints and ceramics is possible without any difficulty. Of great importance for art history studies are tests of the authenticity of relics. Often the shape of objects does not give any clue to the period of origin and relevant conclusions are then drawn from the chemical composition. Not only true historic finds, but sometimes also purposely faked pieces are subjected to such authenticity tests.

Another of the archaeologist's tasks is investigating the connection between various finds of similar shape and elucidation of inaccurate information on the localities of discoveries.

Which instrumental outfit and which one of the procedures of laser micro-spectrochemical analysis are best suited for solving the described problems of archaeology? A non-switched solid-state laser with a pulse output energy of 1 J is fully adequate, such as the ZFL 750 model with a neodymium glass rod. The use of a divergence diaphragm is apparently not necessary. With the model LMA of VEB Carl Zeiss, Jena, focusing of the laser radiation with the reflecting objective $\times 40/0{\cdot}5$ is recommended for most cases, as for instance in the analysis of the paints of old pictures.

The spectrograph for a barium analysis should also be the model Q 24, or a similar medium quartz prism spectrograph, because archaeological investigations, as experience has shown, call for complete evaluation of the heavy metals. A Q-switched laser can have certain advantages in the examination of glazes.

Petrakiew *et al.*[99] analysed with the LMA 1 laser probe the pigments in several layers of old murals, one painted over the other. The purpose of this work was to throw light on the methods and the materials used at the time of origin and to give guidance for the right approach to successful restoration. The pigments were derived from minerals in the form of metal oxides, hydroxides and salts. So far spectrochemical analyses, carried out with arc discharge and cupped carbon electrodes, only served for finding average concentrations. It was not possible to examine the different layers of such paintings individually. By micro-spot-analysis the single layers and the matrix material could now be identified and the polarizing unit and the quarter-wave plate helped to trace layers with low

TABLE 17.[99]

| *Pigment* | *Colour* | *Chemical composition* | *Principal analytical lines used for identification (Å)* | *Specific chemical determination* |
|---|---|---|---|---|
| Azurite | Blue | $2CuCO_3.Cu(OH)_2$ | Cu 3933·7<br>Cu 3968·5 | $Cu^{2+}$ |
| Ultramarine | Blue | $3Na_2O_3.3Al_2O_3$ | Al 3092·7<br>Al 3961·5 | $Al^{3+}$ |
| | | $6SiO_2.2Na_2S$ | Al 3944<br>Al 3082·1 | $Al^{3+}$ |
| Minium (red lead) | Red | $Pb_3O_4$ | Pb 2393·7<br>Pb 2802·1<br>Pb 2823·2<br>Pb 2833<br>Pb 2873·3 | $Pb^{2+}$ |
| Cinnabar | Red | HgS | Hg 4358·3<br>Hg 4046·5<br>Hg 2536·5 | $Hg^{2+}$, $S^{2-}$ |
| Sienna | Brown | $Fe_2O_3.H_2O$ | Fe 4404·7<br>Fe 4383·5<br>Fe 4325·7<br>Fe 4307·9<br>Fe 4271·7 | $Fe^{3+}$ |
| Umber | Brown | $Fe_2O_3.MnO_2$ | Mn 2794·8<br>Mn 2798·2<br>Mn 2576·1<br>Mn 2593·7 | $Fe^{3+}$<br>$Mn^{2+}$ |
| White lead | White | $2PbCO_3.Pb(OH)_2$ | Pb 2393·7<br>Pb 2802·1<br>Pb 2823·2 | $Pb^{2+}$ |
| Bone black | Black | C + P | P 2554·9<br>P 2553·3<br>P 2535·65<br>P 2534·01 | $PO_4^{3-}$ |
| Malachite | Green | $CuCO_3.Cu(OH)_2$ | Cu 3933·7<br>Cu 3968·5 | $Cu^{2+}$ |

contrast. For an analysis at right angles to the surface a cross-section was cut out and embedded in resin.

As can be seen from Tables 17 and 18 published with the report[99] numerous items of different mural paintings in historic buildings in Bulgaria were identified. Lists were made and records assembled so that now valuable conclusions can be drawn regarding the age of the relics and the art schools from which they sprang.

TABLE 18.[99]

| *No.* | *Object* | *Number of layers* | *Colour of examined layer* | *Sample preparation* | *Main elements detected* | *Conclusion* |
|---|---|---|---|---|---|---|
| 1 | Church of St. Peter and St. Paul, Tirnowo | 2 | Blue | In polyester | Cu, Fe, Al, Mg, Ca | Azurite |
| 2 | | 2 | Bright red | In polyester | Hg, Pb, Fe, Ca, Al, Mg | Cinnabar |
| 3 | | 2 | Dark red | In polyester | Hg, Pb, Fe, Al, Mg, Ca | Mixture of cinnabar and minium (red lead) |
| 4 | Kremikow Monastery | 2 | Bright red (crystal) | None | Hg, Fe, Pb, Al, Ca, Mg, Cu | Cinnabar |
| 5 | | 2 | Dark red | None | Fe, Ca, Mg, Al, Cu, Pb, Zn | Red ochre |
| 6 | Thrakish Cemetery in Village of Myglisch | 2 | Red | In polyester | Fe, Ca, Mg, Al, Cu, Ti, Zn | Red ochre |
| 7 | | 1 | Yellow | In polyester | Fe, Ca, Mg, Cu, Al, Ba, Zn | Yellow ochre |

## 4.1.5. Biology and medicine

In 1963 Rosan, Healy and McNary of the Departments of Biochemistry, Pathology and Anatomy of the Medical Center of Boston University, Massachusetts, undertook investigations with the laser microprobe of the Jarrell-Ash Co.[108] They checked in the spectrum region from 255 to 398 nm the elements Mn, Fe, Be, Cu, Zn and Ca in different tracts of the brain of a rat, in islets, acini and the septum of the pancreas, in elastin and in concrements. The work was carried out on tissue sections, on unprepared samples, not contained in any way (calculus), and on solutions dispersed in gels.[106] The report is only of a general nature. It, however, confirms that qualitative and sometimes also quantitative chemical micro-determinations of the contents of biological materials, the latter with an error margin of $\pm 20$ per cent, are feasible and promising.

The first reports have already been published of analytical work with a laser micro-analyser on a living body.[55] An experiment was carried out on an anaesthetized mouse, placed on the sample stage of the microprobe, for finding the distribution of elements in the ear. According to the report it was found that aluminium, copper, silicon and titanium are in a typical way linked with calcium in the process of bone growth.

After irradiation of an index finger with a focused laser light beam (against the safety regulations)[52] changes in the Na/Ca ratio were detected.

Rosan *et al.*[107] found differences in the distribution of Mg, Al, Fe and Zn in various parts of filled teeth. A qualitative analysis of individual particles of a dental alloy is also reported in the literature.[103] The first series of semi-quantitative tests on teeth, stone formations and bones[46, 119] revealed that differences of concentrations of P, Mg, Si, Al and Cu can be more reliably evaluated than those of Fe, Ti and Zn because the quantities of these elements often are below the detection limits ($<0{\cdot}001$ per cent). The histological distribution of Ca and Mg in stomachs of dogs and humans has been quantitatively established with a laser probe.[41, 44]

Japanese research workers[60] found a link between the Mn and Zn concentration and the location of cancer formation within a prostate gland.

As early as 1934 Gerlach[41] pointed out that conventional emission spectrochemical analysis can be applied to great advantage in medical practice and research. In a lecture given in 1939 at Jena during a course for doctors and biologists[42] he again remarked that the great possibilities offered by this method of analysis were only a little exploited. He then called for efforts to establish biochemical distribution laws. Now, a few decades on, we are in the position to report a considerable advance in scientific instrumentation and to offer to biological and medical institutes a micro-method for systematic examination of the distribution of elements in small zones.

Typical for the developments in the methods of histo-chemistry[44] are the change to Q-switched lasers and to photoelectric recording of spectra with powerful grating spectrographs, and also the adoption of internal standards (Ag content of films supporting the vaporized sample material).

The noteworthy suggestion[45] has been made that, because of the very small quantities required, samples of blood and other body fluids should be sent by air mail to central laboratories equipped with a laser micro-spectrum analyser. A case of the successful application of laser micro-analysis before the histo-chemical routine examination proper of skin samples is reported in the literature (Calcinosis Cutis).[133]

The use of the new method suggests itself for diagnosing tissue diseases and metal poisoning through aggregations of Pb, Ag, Bi, Cd, Zn,

Hg and Tl in certain parts of human or animal bodies, for controlling the use of medicines containing heavy metals (for instance in cases of argyrosis) and for the investigation of silicosis. The new technique is also of interest for the work of expert witnesses before law courts and in insurance cases, for instance in questions of the effects of injuries through shooting, war injuries or damage to the eye by foreign bodies (detection of traces of the bullet metal in the wound or analysis of fragments removed by operation – see also § 4.1.6).

For many years scientists have been searching for a solution of the problem of silicosis. Rock and powder samples, the latter mounted on a carrier slide with Canada balsam, were up till now examined optically, mainly in mineral determinations by microscope and by chemical methods in general tests or in investigations of isolated components. The evaluation of the results of these two separately performed test procedures was rarely exact and satisfactory. Microscopic testing of drillings is rather difficult even for an experienced analyst. Subjective errors can never be completely avoided. The recommended combination of microscope and spectroscopic outfit gives reliable results and provides a permanent record of them. The distribution of the particle size of dust suspended in the air depends on the distance from the source. After some drilling in graywacke slate, particles of sizes between 2 and 175 μm were found at 4 metres distance from the face; of these 10 per cent were larger than 100 μm, 50 per cent were between 50 and 100 μm, 20 per cent between 20 and 50 μm and only 15 per cent were smaller than 20 μm.[66a] Particles of this size can easily be examined with a laser micro-analyser. This form of analysis is also possible when the particle diameter is less than 40 μm because pure Canada balsam or a similar material used for embedding does not produce an interfering spectrum.

Similarly deposits can also be detected in lung tissue in cases of silicosis, asbestosis and talcosis simultaneously with microscopic examination. Furthermore the analytical determination of Fe appears to be feasible in the case of siderosis, and also of Ba in barytosis, of Sb and Al in antimony or aluminium lungs, of Sn in stannosis and W, Ti, Co, Ta in so-called hard-metal lungs.

In the workers health service occupational diseases caused by the elements Pb, Hg, As, P, Mn, Cd, Cr, Be and V are well known. It may become necessary in a diagnosis to differentiate between Pb and Bi. The lead line of a patients gum edge cannot be distinguished from a bismuth line. The bismuth might derive from a bismuth treatment for lues. Lead poisoning or its absence can also be spectrochemically proved. Furthermore cadmium can be detected in teeth, vanadium in a green lining of the tongue and chromium in skin ulcers.

On the whole a great amount of work still remains to be done in the study of the distribution of elements in the human and animal body and in plants. The metabolism of metals and their accumulation in certain organs or plant tissues has first to be understood to find quickly the causes of disease, death and injuries.

The introduction of the electron probe into medicine also has to be mentioned. Mellors and Carroll[77] for instance experimented with 4 μm microtome tissue sections and concerned themselves especially with the iron distribution in the human synovial membrane and with the iron and chromium concentration in joint capsules.

The large instruments of electron probe outfits are of course available for doctors only in research centres. But the use of laser micro-spectrum analysers in smaller laboratories in the not so distant future can be foreseen. This is also indicated by the latest successes of scientists at Stanford University, California.[10] They detected quantities as small as $10^{-14}$ to $10^{-16}$ g of Ca, Mg, Zn, Cu and Mn in frozen sections of human liver and kidney, about $10^{-13}$ g Fe in individual red blood corpuscles and $10^{-15}$ g Zn in the sperm of rats.

### 4.1.6. Crime detection and forensic medicine

Departments responsible for the safety of the state and its citizens and for the detection of crimes and offences depend on the active collaboration of well-equipped laboratories. Emission and absorption spectroscopy of all wavelengths has for decades been established, continuously improved and developed in analytical institutes. The new method of micro-emission spectroscopy with sample vaporization by laser radiation seems to offer, particularly in this field, some significant advantages:

1 The destruction-free chemical analysis of an object providing evidence is very valuable because, if required, the object is then still available for further investigation. If a spectrograph with a large working range like the Q 24 type is used, up to 60 elements can be determined in one operation on a single photographic plate. The chemical analysis need not be the final step in a test program.
2 Often probably only small objects or minute quantities of material are at the disposal of the investigators, such as paint, lacquer or soil traces, combustion residues, metal traces at the point of impact of a bullet, splinters, chips, fragments of metal, glass, ceramics, stones etc.
3 A really rapid method of analysis is very desirable.
4 Also desirable is the photographic recording of the results as documentation.

The type of test samples met with in crime detection and the elements or element groups in question may be any of those dealt with in the foregoing § 4.1.1. to § 4.1.5. and in the following under § 4.2.1. to § 4.2.14.

In the discussion of the application of laser micro-analysis to archaeology the exposing of forgeries has already been mentioned. Similarly, what has been said before about identifications and determination of origins also applies here; these problems are not unique to archaeology.

Among the subjects of possible interest in forensic medicine are injuries by shooting, which are of special importance in cases of murder, suicide or accident. The main concern here is the reconstruction of the course of events and with it the establishing of the cause of death according to forensic medicine. Some of the most important questions can be answered with the help of spectrochemical analysis, for instance: Where did the bullet enter? What was the path of the projectile? What weapon was used? The contamination around the hole of entry of the bullet into the body or the clothing is examined. The missile carries rust, dirt or oil particles from the weapon with it and deposits metal traces (for example Pb, Sb, Fe, Cu, Hg) and powder smoke residue.

Fig. 85 shows the point of entry of a bullet in shirt material. At the order of a forensic institute the authors examined with the LMA 1 laser micro-spectrum analyser the visible irregular deposits and the small fusion

Area analysed with the LMA outfit

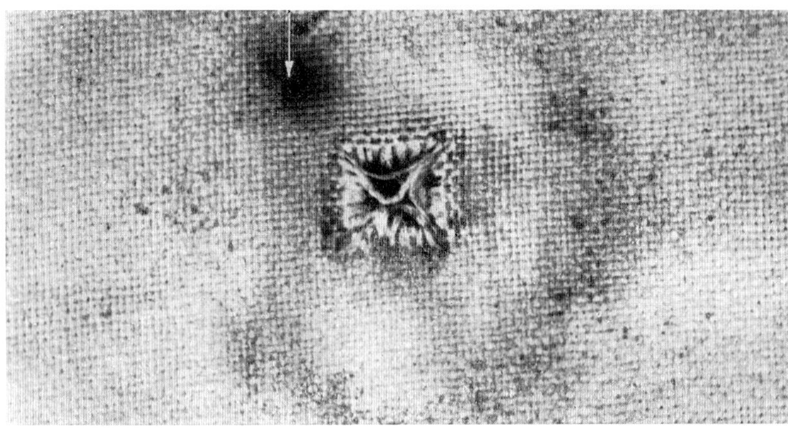

FIG. 85. Application of the laser micro-spectrum analyser in crime investigations: hole of bullet entry in shirt with powder smoke and metal traces.

beads. The main constituents of the former were found to be Pb, Sb, and Ca, besides traces of Mg, Cu, Sn, Ag, Si and Al. The fusion beads of about 10 μm contained Pb in addition to small traces of Mg.

In the case of road accidents it is important to obtain clues, for instance by investigating glass, paint (Table 19) or metal traces in wounds. On several occasions the authors analysed spectroscopically paint chips from cars to establish the possible presence of wood and metal particles or textile shreads. Various elements can be determined in such samples with five laser pulses. In one particular case the different concentrations of a number of heavy metals carried special weight in the evidence. These metals were found besides Al, Ca, Cu, Fe, Mg, Si and a great amount of Ti,

as listed (Table 19). Neuninger[93] examined the different layers of car paint samples for identification purposes in cases of road accidents in which the car involved had driven off. For his tests paint chips were embedded in casting resin and a cross-section, at right angles to the surface, was prepared and polished. Neuninger comments:
'Gerade hier zeigt sich die deutliche Überlegenheit der Lasermikrospektralanalyse gegenüber einer Untersuchung mit der Elektronenstrahl-Mikrosonde. Zunächst müßte dort der Schliff bedampft werden (z.B. mit Kohlenstoff oder zweckmäßiger mit Gold), um die Oberfläche elektrisch leitend zu machen. Das Bedampfen erschwert aber die genaue Einstellung einer bestimmten Schicht unter dem Mikroskop, da eine sichere Differenzierung der einzelnen Schichten infolge der geringen Durchsichtigkeit der Bedampfungsschicht schwierig ist. Die hohe Brennflecktemperatur des Elektronenstrahles erhitzt die organische Matrix so stark, daß sie schmilzt und teilweise sogar verkohlt. Eine genaue Bestimmung der Zusammensetzung einzelner Lackschichten ist daher, wie eigene Versuche ergaben, mit der Elektronenstrahl-Mikrosonde nicht möglich.' [Here in particular the clear advantage of laser miro-spectrochemical analysis over the electron beam microprobe method becomes evident. For the latter the polished surface has to be coated, for example with carbon or better with gold, to render it electrically conducting. Such a coating makes exact focusing of the microscope on one individual layer difficult because, as a result of its low transparency, the various layers cannot easily be distinguished. The high temperature at the focal spot of the electron beam heats the organic matrix to such a degree that it melts or sometimes is even carbonized. Specific experiments have shown that the composition of individual paint layers cannot accurately be determined with an electron beam microprobe.]

TABLE 19. Distribution of several elements in car paint samples

| *Number of sample* | *Colour* | *Elements* | | | | |
|---|---|---|---|---|---|---|
| | | *As* | *Co* | *Cr* | *Mn* | *Pb* |
| 1 | Red | – | + | – | tr | tr |
| 2 | Yellow | tr | – | ++ | + | + |
| 3 | Green | – | tr | tr | – | + |

Symbols: – not detectable tr trace element
\+ ++ Concentration ranges of main constituents

In solving a crime of fraud it could be proved with the LMA 1 outfit that gold watches set with brilliants were of inferior quality because under a fairly thick layer of gold bodies of brass were hidden. An important feature of the method is that no effect of the analysis could be seen on the samples with the naked eye; and these samples would have been of considerable value had the test result been different.

After the burglary of a safe that had been forced with an acetylene cutter Neuninger in search of evidence went over the clothing of a suspect with a vacuum cleaner and then looked for fusion beads. He succeeded in proving that the globules he found were of the same material as the safe.

So far the analysis of minute glass pieces, for instance fragments of car head-light screens, has always caused difficulties for the spectroscopists. Now laser micro-analysis gives reliable results for coloured glasses, glass pieces with rough surfaces, and also for transparent glass samples smoked in a small benzene flame.[93]

The detection limits found in the quantitative analysis of minerals are given in Table 22.[93]

E. Rudolph[111] was successful in his attempt at applying the laser micro-spectrum analyser LMA 1 to the determination of foreign metals rubbed off on steel tools. The elements of particular interest were Cu, Al and Pb and their alloys; they could reliably be analysed on tools used for crimes on which they were rubbed off in point-like or more extended areas of contact. The scouring need not necessarily be visible with the naked eye or under the microscope before the analysis; scanning of the suspected part of the surface produced the required result. Here it was found to be an advantage that through appropriate choice of the working parameters the iron as the main constituent of the steel tools was only slightly excited; this facilitated the detection of the other foreign elements.

Rudolph in his work on metal scouring included abrasion in a wider sense, as for instance the transfer of material in short circuits. These crime investigations were, furthermore, concerned with the analysis of protecting coverings and coatings and dirt deposits. Neuninger[93] analysed copper traces rubbed off on the penknife of a man suspected of the theft of copper cables.

What has been said in §4.1.5 about occupational diseases also applies to the problems of poisoning. Poisoning by inorganic substances, intentional or accidental, is comparatively rare, apart from arsenic poisoning. One gramme of thallium is considered a fatal dose; concentration of this element in bones and in hair is a well-known phenomenon. The authors have analysed individual hairs from the head of a person showing poisoning symptoms. A laser beam can be very accurately concentrated on the root or the shaft because the crater diameter is always smaller than the diameter of a hair of the head. According to preliminary tests the various parts of the hair contain different quantities of Ca and Mg. The distribution of elements in hair also varies remarkably from person to person.

Mees's stripe is a grey band, 1 mm wide, that appears after some time on the nail plates; it is sometimes also observed in cases of arsenic poisoning.

Whatever the type of sample, whether metallic or non-metallic, natural or artificially produced, or medical-biological, the appropriate sections of this monograph on laser micro-spectrochemical analysis should always be referred to because repetitions had to be avoided.

## 4.2. THE MICRO-ANALYSIS OF PARTICULAR ELEMENTS AND ELEMENT GROUPS

The treatment in this section of the analysis of different chemical elements is based on the experiences gained at the VEB Carl Zeiss, Jena.

### 4.2.1. Copper – silver – gold

The main analytical spectrum lines of these elements of the first subgroup of the periodic system lie in the ultra-violet; they are without exception also used in laser micro-spectrochemical analysis. They can satisfactorily be detected with a quartz prism spectrograph of the type of the Q 24 model of VEB Carl Zeiss, Jena (Fig. 86); to this type also belong the models ISP-28 from the USSR and E 498 of Hilger & Watts Ltd, London (now the Analytical Division of Rank Precision Industries Ltd) and furthermore an instrument of the Japanese firm of Shimadzu Seisakusho Ltd. of Kyoto and the spectrograph of the firm of Fuess, Berlin–Steglitz.

For the determination of copper in ores and light metal alloys like nickel silver and brass, the Cu I lines 327·396 and 324·754 nm as well as a number of Cu II lines are available. The well-known lines at 224·261 and 224·699 nm of the latter group are, because of self-absorption, not suitable for quantitative copper analysis with a laser probe. In laser micro-spectroscopy the degree of self-absorption of Cu II and Cu I lines depends very much on the nature of the test substance. In the spectra of copper iron sulphides the line Cu I 324·754 nm exhibits very strong self-absorption and the line Cu I 327·296 nm hardly any at all. The authors found that self-absorption is considerably stronger in carbonate than in sulphide spectra. The Cu II lines 227·625, 229·436, 236·989 and 240·011 nm are free from interfering self-absorption.

The two lines primarily to be considered for silver detection are the Ag I lines 338·289 and 338·068 nm. In checking, especially of silver ores, one can also resort to Ag II lines with an excitation energy between 10 and 11 eV, such as: Ag II 267·752, 266·046, 253·531 (9·93 eV), 250·660, 247·380, 243·779 (9·93 eV), 241·318 and 241·135 nm. Of these, the line of the shortest wavelength that can be recorded in the argentite spectrum on an appropriate photographic plate, for example the ORWO WU 3 spectrum plate (blue-extra hard),is Ag II 211·383 nm. For a rapid analysis only the most sensitive Ag I lines above 338 nm are of practical value because the mentioned Ag II lines do not occur in the spectrum tables for regular routine tests.

The determination of gold in Au–Sb alloys and in coins with vaporization by laser radiation can be based on the Au I lines 312·278, 267·595 and 242·795 nm; no self-absorption was observed. What has been said so far resulted from experience in the method with the spark gap between the sample and the microscope objective.

If the microplasma is directly imaged on the spectrograph slit the spectrum resulting on the photographic plate, according to Debras–Guédon and Liodec,[27] contains Cu lines, but no Ag lines. This is a significant

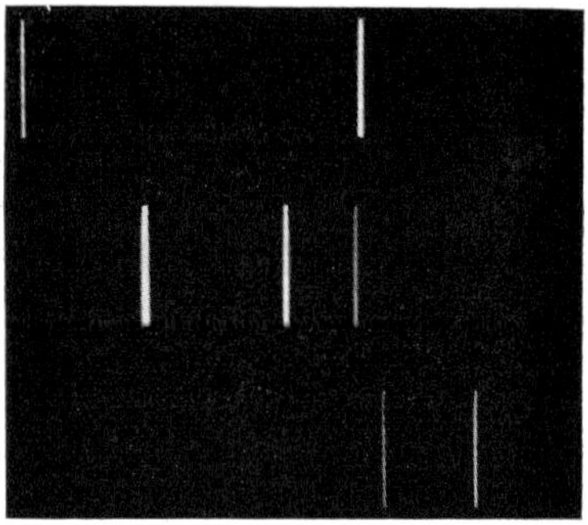

(a)

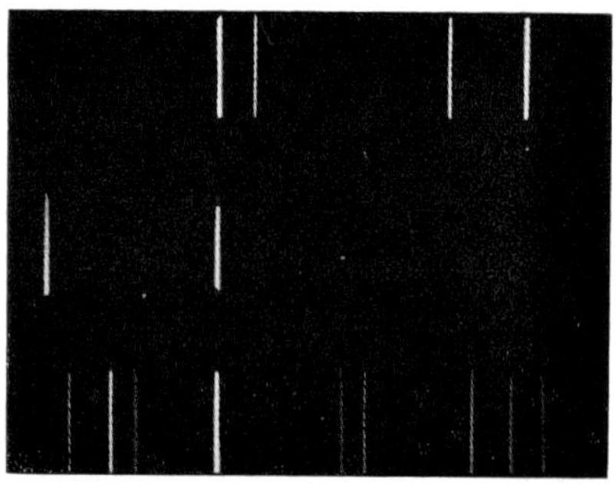

(b)

FIG. 86. Sulphide ore analysis.

(*a*) Identification of Ag, Zn and Cu with arc lines:
Top: Ag I 328·07, 338·29 nm.
Middle: Zn 328·23, 330·26, 334·50 nm.
Bottom: Cu I 327·4, 324·75 nm.
(*b*) Identification of Ag, Zn and Fe with spark lines:
Top: Ag II 243·78, 244·79, 247·38 nm.
Middle: Zn II 250·2 nm.
Bottom: Fe II 243·93, 244·45, 245·9, 249·3, 249·78 nm.
All three spectrum sections show the C I line 247·86 nm due to the electrodes.

finding that makes it inadvisable at the present state of development of experimental technique to work in laser micro-spectroscopy without cross-excitation.

The detection posibility for silver and copper is in laser micro-analysis greater than for gold and other elements. But the surface of a sample can very easily be contaminated with these heavy metals if the most exacting cleanliness is not observed in handling tools and photochemical materials.

Silver traces in lead sulphides were detected with Q-switched and with non-switched solid-state lasers, in the latter case both with ruby and with neodymium glass resonators. An absolute detection limit of less than $10^{-8}$ g was established. In 1969 in Tokyo limits of $10^{-13}$ g for silver and copper were found.[73] Practical experience of copper determination was gained with copper iron sulphides (chalcopyrite), copper germanium iron sulphides (germanite), carbonates (malachite, azurite) and alloys of the nickel silver and brass type.

Silver can be determined in small crystals of argentite, proustite and polybasite with one pulse from a solid-state laser with an output energy of about 0·4 J.

### 4.2.2. Zinc – cadmium – mercury

These elements are extremely volatile; for their spectroscopic determination by the conventional emission method an appropriate shape and depth of the bore holes of the auxiliary electrodes must be determined with care. An unsteady arc discharge, frequently observed at the beginning of the vaporization, can lead to completely wrong conclusions on presence or absence of one or more of these elements in the second subgroup of the periodic system.

Such analytical difficulties dreaded by spectroscopists do not arise in the laser micro-procedure because the excitation process only lasts for a period of the order of milliseconds and fractionation cannot develop. Zn, Cd and Hg can be sufficiently excited in the spark gap between the sample surface and the protecting plate in front of an objective with long working distance. French scientists using the laser method without auxiliary excitation only succeeded in detecting zinc with the help of the Zn I line 334·502 nm; cadmium and mercury could not be found in the microplasma emanating from samples rich in these elements.

In the two-step procedure of laser micro-analysis with carbon electrodes other arc lines besides the Zn I line just mentioned, namely 330·259 and 328·233 nm, can serve for the evaluation even if no inductance is inserted in the circuit of the mains supply unit. There are useful Zn II lines still further in ultra-violet, namely at 255·796 and 250·200 nm, both with an excitation energy of 10·95 eV.

In the spectrum of sphalerite taken with the spectrograph model Q 24 on ORWO photographic plate type WU-3, the Zn I line 213·856 nm shows very strong self-absorption; it is hardly of any analytical value.

Besides analytical lines in the ultra-violet there are also suitable spectrum lines of cadmium and mercury in the visible spectrum region. Ultra-violet Cd I lines for low concentrations are at 361·051, 346·766 and 340·365 nm; the intensity and degree of self-absorption decrease in this order. Other lines for the analysis of ores and alloys are: Cd II 274·858, 257·309, 232·115 and 231·284 nm.

Mercury can also be determined with blue-sensitive spectrum plates by means of the lines Hg I 435·835 and 404·656 nm, without interference by cyanogen bands. Further Hg I lines within the working range of the spectrograph model Q 24 proved to be of value in practical work, are: 365·015, doublet 313·183/312·566, 296·728, 284·783 and 253·652 nm. The line Hg II 222·471 nm can serve for checking.

The Hg I line 366·328 nm is emphazised in spectrum line tables and marked in comparison spectrum charts. The authors found that this line is hardly visible even in the cinnabar spectrum which, when taken with a 15 μm wide slit and examined under a magnification of × 20, shows lines of millimetre widths.

### 4.2.3. Scandium – yttrium – rare earths

Spectrum line tables and comparison spectrum charts list the Sc II lines 361·38 and 363·07 nm as suitable analytical lines. As the reproduction (Fig. 87) of the relevant section of a thortveitite spectrum shows, there are altogether eleven Sc lines found here when analysing scandium silicates (and also oxides). Hence such an investigation with a laser probe can be restricted to the region of the mentioned lines. It has to be noted that this is not possible when non-conducting substances are excited in the conventional way by an arc between carbon or graphite electrodes because particularly in this spectrum region band systems often interfere. In routine work the authors have often determined Sc, Y and La in minerals with a laser micro-spectrum analyser and carbon auxiliary electrodes and photographed the spectra obtained with a medium quartz prism spectrograph type Q 24.

Whereas in scandium determinations various arc lines in the spectrum region above 390 nm may be employed, one is restricted to Y II lines in detecting main constituents and traces of yttrium in solid samples. As a result of the very spark-line nature of the spectroscopic light source in the two-step excitation procedure this is no disadvantage.

In Fig. 87 three of these spectrum lines can be seen which were emitted by a samarskite sample: Y II 371·03, 377·4 and 378·87 nm. Oxides and silicates, for instance orthite, emit the line Y II 437·49 nm with the same intensity as Y II 371·03 nm. The main analytical line of yttrium is at 360·07 nm; yet the authors only used it for trace determinations because the first mentioned lines could always be well identified in higher yttrium concentrations of several per cent.

An example of a successful yttrium determination is the detection of traces in the titanite from the syenite occurring in the Plauen district near Dresden. The authors furthermore obtained, with two or three laser pul-

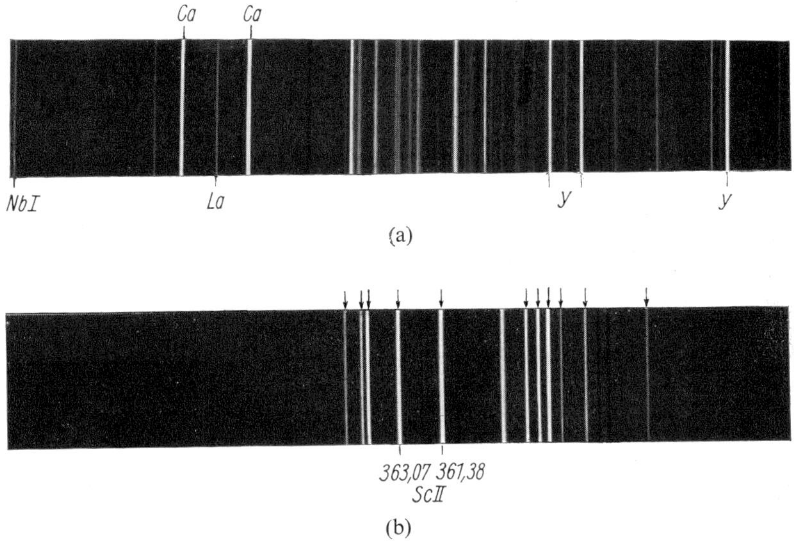


FIG. 87. Identification of elements of the third subgroup of the periodic system in the visible spectrum region of the working range of a medium quartz prism spectrograph (model Q 24).

(*a*) La and Y lines in the spectrum of the oxide mineral samarskite. Between the Ca II lines 393·37 and 396·85 nm the line La II 394·91 nm can be seen, and to the right the Y II lines 378·87, 377·4 and 371·03 nm. The line Nb I 405·89 nm is marked at the left end of the spectrum section.

(*b*) A series of eleven Sc lines in the thortveitite spectrum with the more familiar analytical lines Sc II 361·38 and 363·07 nm.

ses, evidence of yttrium in the scandium silicate thortveitite and in the cerium phosphate monazite.

Lanthanum can be identified with a grating or three-prism glass spectrograph using La I lines, mainly La I 624·99 nm, or with a quartz prism spectrograph using La II lines. The authors decided on the spectrograph Q 24 and La II lines as more practical. It is fairly easy to find the main analytical line La II 394·91 nm because oxides and silicates rich in lanthanum always contain calcium in addition; on either side of this La II line and close to it lie two intense Ca II lines at 393·37 and 396·85 nm. In the spectra of compounds with low aluminium content (for example orthite, see Fig. 64) also the aluminium lines Al I 394·4 and 396·15 nm can be used as reference for reliable identification of the La line.

Further sensitive La II lines of about equal intensities lie at 407·7, 408·67, 412·3 and 433·37 nm; they can still be found with ORWO photographic plates WU-3, especially in specimens of samarskite, orthite, tritomite and cerite.

When cerium is to be determined with a Q 24 type spectrograph in samples containing calcium the Ce II line 422·26 nm cannot be used; instead the line Ce II 418·66 nm is available, whereas the line Ce II 413·38 nm only shows up in the spectrum of sample parts with high cerium concentration. Quite generally it must be said that, even with a laser probe, cerium, not having intense analytical lines, can only be detected in samples of a specific cerium mineral or in particular sample regions with specially high cerium concentration.

Conditions are completely different for determining ytterbium with a laser micro-spectrum analyser; for this element several intense analytical lines are available, namely Yb II 328·9 and 369·4 nm as well as Yb I 398·8 nm; these lines were found by the authors in the silicate spectra of orthite, tritomite, cerite and titanite with a neodymium glass laser emitting radiation of wavelength 1060 nm with an output energy of 0·5 J.

The authors detected praseodymium, neodymium, samarium, gadolinium, dysprosium and erbium in their oxides with crossexcitation in the two-step procedure, using the following main analytical spectrum lines:

| *Element* | *Analytical lines* |
|---|---|
| Praseodymium | Pr 422·5/422·3, 417·9, 248·87 nm |
| Neodymium | Nd II 401·2, 406·1, 410·9 nm |
| Samarium | Sm II 442·05/442·4/443·4 nm |
| Gadolinium | Gd II 335·05, 335·86, 336·2 nm |
| Dysprosium | Dy II 400·04, 407·79 nm |
| Erbium | Er 349·9, 369·3, 393·9 nm |
| Holmium | Ho II 293·68 nm |

Holmium was found as a minor constituent in dysprosium oxide.

### 4.2.4. Titanium – zirconium – hafnium – thorium

Because the strongest Ti I analytical lines lie in the visible above about 498 nm, Ti II lines are chiefly to be considered for work with a laser micro-spectrum analyser and a quartz prism spectrograph. The authors relied mainly on evaluation of the triplet Ti II 334·94, 336·12 and 337·28 nm, as shown in the section of an ilmenite ($FeTiO_3$) spectrum in Fig. 88(a). On occasions, especially when Ca, La and Ti cations together were to be detected in the long wavelength section of the Q 24 spectrograph working range, it appeared advisable to resort also to the Ti II lines 368·52, 375·93 and 376·13 nm.

It is not difficult with a solid-state laser of 0·5 J output energy to detect titanium in small crystals of anatase, ilmenite, perovskite, titanite or melanite.

Tests with a Q-switched ruby laser proved that titanium can be found as a trace element in chromite, in pyroxenes andin potassium felspar, and as a major component of enstatite. Ti II lines, namely 334·94, 337·28 and 338·38 nm, were also used for determining titanium in meteorite stones with a Q-switched ruby laser.

The Ti II line 348·38 nm has been found in the microplasma over oxide samples. In checking the emission spectra of the silicates zircon and tritomite the authors established that the group of lines Zr II 342·8, 346·3, 347·94/348·1 and 349·6 nm emitted by silicate samples is very suitable for reliable determination of zirconium. In addition the physically strongest line Zr II 339·19 nm is available. One also has to remember that zirconium has an extremely low volatility and therefore often is very difficult to detect by conventional emission spectroscopy. When vaporizing specimens with high zirconium content by means of a neodymium glass resonator the authors always obtained regular circular craters and there was no indication of sample spattering.

Italian workers[24] based their study of Zircaloy compounds on the line Zr 345·75 nm.

Hafnium is one of the elements the analytical lines of which are not included even in the spectrum charts most widely used in practical work. Geochemists often resort to the line Hf 307·3 nm for establishing Hf; this, however, is an arc line and it is therefore better to consider the line Hf 277·3 nm or any other Hf II line.

Thorium emits the useful analytical lines 374·1 and 297·86 nm which serve for identifying silicates like thorite and titanite and detecting Th in the phosphate monazite, etc.

### 4.2.5. Vanadium – niobium – tantalum

The selection of analytical lines from the many vanadium lines for identifying minute vanadate crystals or for determining vanadium in sorting high speed steels can essentially be restricted to well tried V II lines; this can be seen from Fig. 88(b) in which four such lines are marked, namely 309·3, 310·23, 311·07 and 311·84 nm.

One can, however, also employ V I lines for detecting excited levels in microplasmas produced by laser irradiation close to the surface of oxides (V I 405·36 nm etc.).[27]

Niobium content, for instance of the oxide samarskite, can satisfactorily be ascertained by means of the Nb I lines 405·89 and 407·97 nm. The

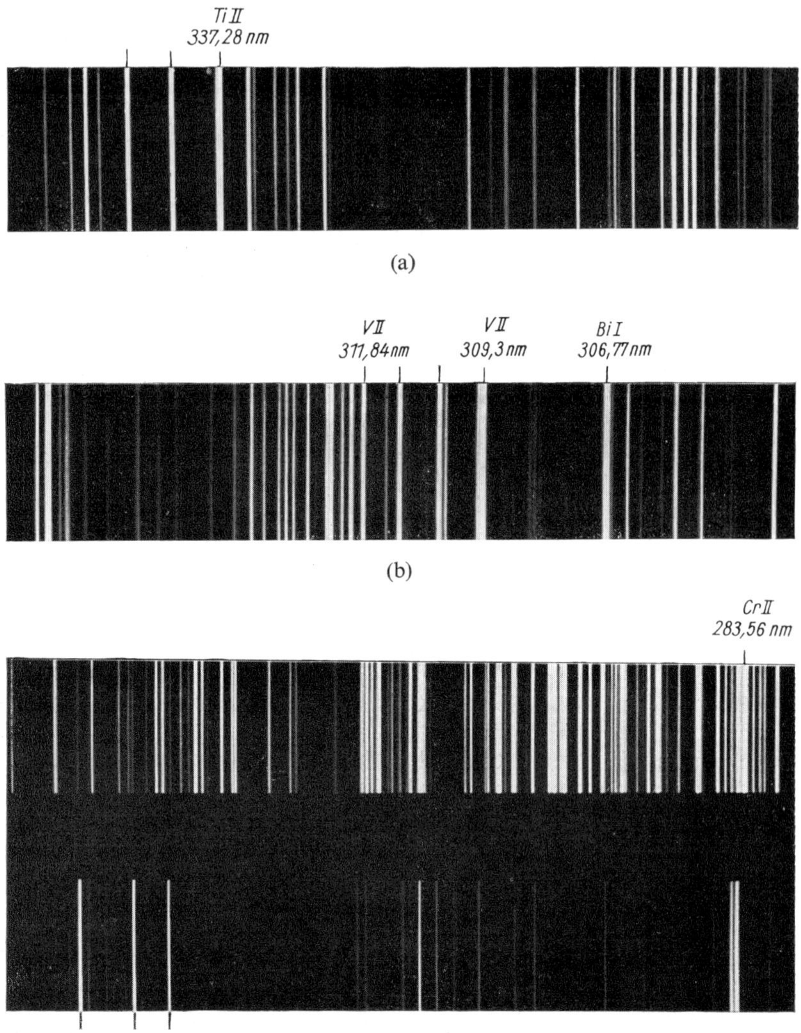


(c)

FIG. 88. Main analytical lines of metal elements important in technology (atomic numbers 22 to 28, 42, 74) and neighbouring lines in mineral emission spectra.

(*a*) Ti II lines 334·94, 336·12, 337·28 nm in the ilmenite ($FeTiO_3$) spectrum.

(*b*) V II lines in the spectrum of the bismuth vanadate pucherite. The following lines are marked (from right to left): V II 309·3, 310·23, 311·07, 311·84 nm.

(*c*) Lines in the emission spectrum of the lead chromate crocoite near the Cr II line 283·56 nm. Underneath (on left): manganese triplet 293·3, 293·9, 294·9 nm in the spectrum of the phosphate triphylite.

CN band 421·6 nm does not appear in the spectra obtained with laser micro-spectrum analysers.

The most sensitive tantalum lines suitable for examining oxides ores and special alloys lie in the ultra-violet, namely Ta I 271·5 and 243·7 nm.

### 4.2.6. Chromium – manganese – iron – cobalt – nickel

Mineral and steel analyses can be based on Cr I lines in the visible and Cr II lines in the ultra-violet spectrum range. The triplet Cr I 425·4, 427·48, 428·97 nm for instance is obtained with one or two laser pulses in the spectra of the oxide chromite and the chromate mineral crocoite. Fig. 88(c) shows a spectrum section including the analytical line Cr II

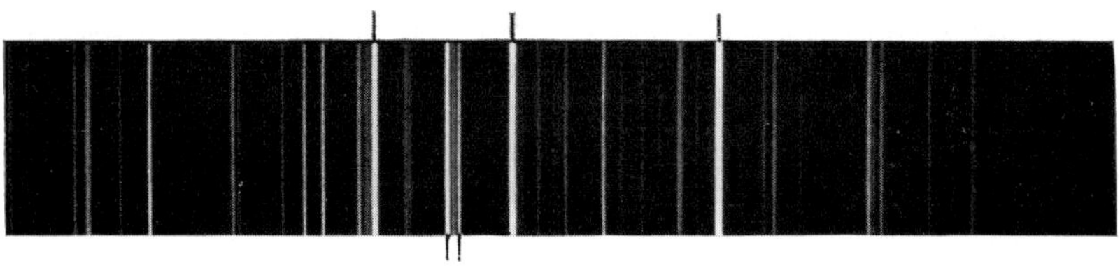

(d)

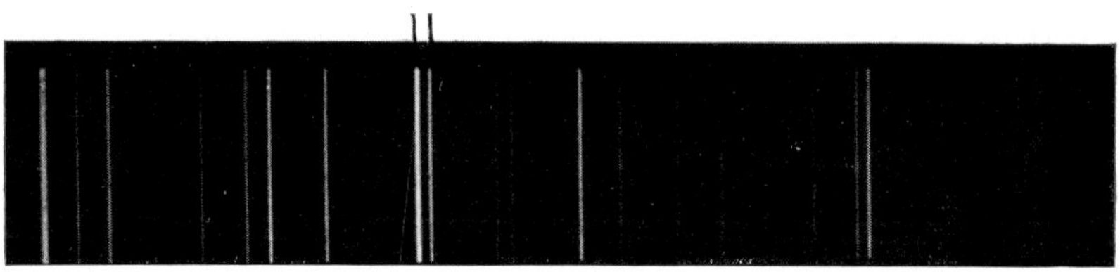

(e)

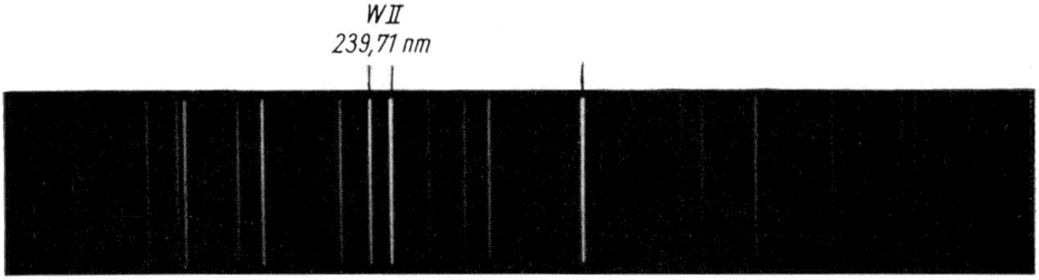


(f)

FIG. 88

(*d*) Mn II triplet 257·6, 259·37, 260·57 nm and the main analytical Fe II lines 259·84 and 259·94 nm in the wolframite spectrum.

(*e*) Fe II line pair 259·84/259·94 nm and neighbouring lines in the pyrite ($FeS_2$) spectrum.

(*f*) Fe II lines 238·2, 239·56 nm and W II line 239·71 nm in thewolframite spectrum.

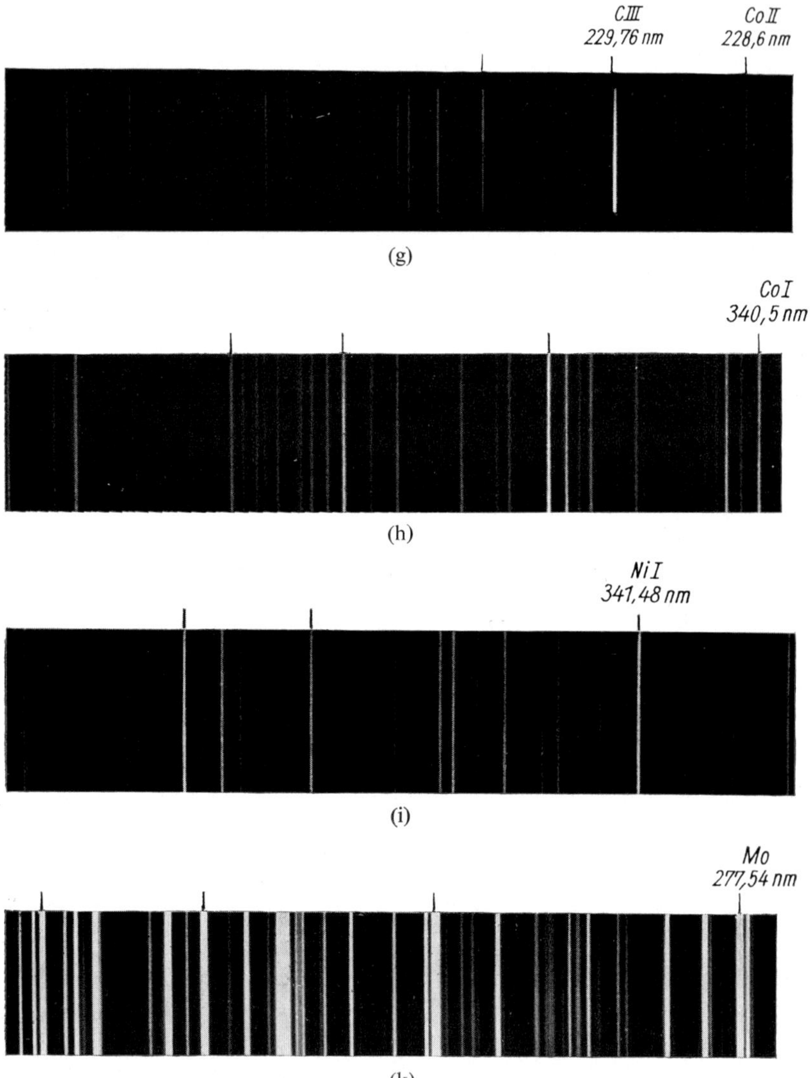


FIG. 88

(*g*) C III line 229·76 nm, with Co II lines 228·6 and 230·79 nm on either side, in the spectrum of the arsenate erythrite.

(*h*) Co I lines 340·5, 345·35, 352·98 nm in the spectrum of the arsenate erythrite.

(*i*) Ni I lines 341·48, 349·3, 352·45 nm and others in the spectrum of the arsenate cabreri-

(*k*) te. Molybdenum analytical lines Mo 277·54 nm, Mo I I281·6, 284·8, 287·15 nm in the molybdenite spectrum.

283·56 nm which can also be evaluated in spectra containing a great number of lines.

Runge, Minck and Bryan[113] found that in the quantitative analysis of steels using a Littrow quartz spectrograph without an auxiliary spark gap one can utilize the line pair Cr 344·56 nm/Fe 344·379 nm.

Vilnat *et al.* in the research laboratory of the Société Française de Céramique, Paris, employed a non-switched ruby laser of 100 J for the quantitative analysis of steels for Cr (0·02 to 2 per cent) and Ni (0·05 to 1 per cent) contents. The standard deviation was 0·04 of a concentration of 0·92 per cent Cr. The still uncompleted investigations are being carried out without cross excitation of the material vaporized by laser irradiation.[129b]

For detection of manganese (according to Matsushita *et al.*[73] the limit with pressed standard powders is $10^{-13}$ g) and determination of the concentration it is advisable to refer to some typical triplets. When the authors had to analyse main constituents of carbonates, tungstates, Li–Fe–Mn phosphates, Mg–Fe–Mn borates or trace elements of silicates, for example Mn in the zinc silicate willemite, they always decided on the groups Mn II 293·3, 293·9, 294·9 nm and Cr II 257·6, 259·37, 260·57 nm.

Iron contents can be suitably determined in laser micro-spectroscopy by means of the ultra-violet line pairs Fe II 259·84/259·94 nm and Fe II 238·2/239·56 nm; this applies to sulphides and oxides as well as to tungstates and silicates.

In the examination of materials containing Co and Ni with a grating spectrograph which only allows coverage of certain sections of the spectrum at a time, Fe I 302·064 nm is a favoured line. Suitable principal analytical lines are then found within a very small extent, such as Ni I 310·155 and 305·08 nm, Co I 340·5 nm.

The iron line Fe 358·1 nm was used in examining the 0·07–0·10 per cent Fe content of the Zr alloy 'Zircaloy'.[23,25]

In the USA the line pair Ni I 341·48 nm/Fe I 344·38 nm was measured photometrically in the quantitative analysis of nickel in steels without cross excitation; the iron line was also used in combination with the Cr line 344·56 nm in finding Cr contents with a Littrow quartz spectrograph.

The use of the iron spectrum with its many lines as wavelength standard in the ultra-violet and visible spectrum region suggests itself also in laser spectroscopy (micro- and macro-procedures); it is less suitable for the red end of the visible and the infra-red regions.

Fig. 88(h) shows further arc lines of cobalt in a section of the spectrum of the arsenate erythrite, namely Co I 345·35, 350·2, 352·98 nm.

As a result of the very spark-like nature of the spectroscopic light source in the two-step excitation method of laser spectrochemical analysis the authors were also able to resort to Co II analytical lines. The lines Co II 228·6 and 230·79 nm are especially easy to find in spectra of alloys and ores on either side of and close to the always present carbon line C III 229·76 nm.

Exactly the same situation is met, as could almost be expected, in the search for nickel, namely: Ni I arc lines in the long-wave ultra-violet region and useful Ni II lines below 290 nm, mainly Ni II 221·65 nm.The section of the cabrerite spectrum illustrated in Fig. 88(i) shows the Ni I analytical lines preferred by the authors namely 341·48, 349·3, 352·45 nm and others.

Whereas the elements Cr, Mn, Fe and Ni could directly be detected in the microplasma produced by laser irradiation on the surface of stainless steel, Co determination was not possible in work without cross excitation.

### 4.2.7. Molybdenum – tungsten – uranium

The spectrum reproduced in Fig. 88(k) shows that near the often recommended analytical Mo II lines 277·54, 281·6, 284·8 and 287·15 nm there are a number of other Mo lines of similar intensities. One is occasionally inclined to omit the look for other elements in the spectra of typical molybdenum minerals which are so very rich in lines. Although the physically strongest arc line Mo I 379·8 nm like the neighbouring lines Mo I 386·4 and 390·3 nm lie in the zone of a CN band system they can be evaluated without difficulty in a spectrum obtained with a laser micro-spectrum analyser.

Molybdenite, the most important molybdenum ore, is often mistaken for graphite. In case of doubt a single laser firing at the minerals frequently found in rocks in coarse scaly aggregates is sufficient for a quick decision without any sample preparation. Although no area of any appreciable extent on the curved flakes of the aggregates can be seen sharply focused under the microscope, no wrong analyses were ever made.

The main W I analytical lines are very weak even in the emission spectrum of wolframite; therefore only W II lines are used in the detection of tungsten. The lines W II 239·71 and 258·9 nm are suitable for the analysis also in the presence of high Fe and Mn contents.

In the spectroscopic determination of uranium one has to content oneself with not very sensitive analytical lines even in laser micro-work. A further difficulty lies in the fact that no positions of uranium lines are entered even in otherwise satisfactory spectrum atlases. As the basic outfit of the LMA Zeiss laser micro-spectrum analyser includes a quartz spectrograph for the ultra-violet the authors preferred lines in this spectrum region when testing oxide and phosphate ores for uranium contents, in particular the lines 237·8, 239·7 and 240·6 nm. The dreaded background continuum emission is considerably weaker with laser excitation than for instance with d.c. continuous arc excitation of samples rich in uranium. The lines U 424·17 and U 424·43 nm can very well be used in work with a grating spectrograph; they cannot of course be resolved with a medium prism spectrograph. In the analysis of nuclear fuels the line U 367·0 nm has been preferred.[24]

### 4.2.8. Lithium – sodium – potassium – rubidium – caesium

When elements of the first main group of the periodic system are to be detected with a laser micro-analyser a powerful glass prism or grating spectrograph is definitely required because nearly all principal analytical lines lie above 580 nm.

The rubidium lines Rb I 780·02 and 794·76 nm and the caesium lines Cs I 894·35 and 852·11 nm have to be recorded with infra-red sensitive plates. The authors have had good experiences with ORWO plates I 850 which have a working range from about 700 to 900 nm.

The following analytical lines can be photographed with ORWO red and infra-red sensitive plates:

| | |
|---|---|
| Lithium: | Wavelengths 670·78 and 610·36 nm |
| Sodium: | D lines |
| Potassium: | Wavelengths 766·49 and 769·89 nm |

In some special analytical cases the authors also detected a few of the alkali metals with the Q 24 type spectrograph. They succeeded with lithium in phosphates, using the line Li I 323·26 nm even when the other cations were Fe and Mn which emit spectra with a considerable number of lines [triphylite ($Li(Fe,Mn)PO_4$)]. It is not practical to determine sodium with the help of the doublet 330·299/330·232 nm; this was brought out by a test on cryolite ($Na_3AlF_6$) when the two main analytical Cu I lines in the ultra-violet, that is the lines for finding a trace element, were of almost exactly the same intensity as the lines of the cation.

An attempt to detect caesium in the tectosilicate pollucite with a 0·4 J laser by means of the line Cs I 455·5 nm was not successful. This however was not only due to low line intensity, but such failures are caused by inadequate vaporization. Some colourless or slightly coloured silicates that are nearly free from iron, manganese and magnesium do not even absorb enough focused laser radiation; and so do certain halides and sulphates. For this reason alone the physically strongest lines in the red and infra-red spectrum regions have to be used. Fundamentally a grating spectrograph could take the place of the suggested glass prism spectrograph such as the very good system of the new three-prism spectrograph of VEB Carl Zeiss, Jena, consisting of a 70 cm collimator, a 70 cm camera and Försterling's prism set. However, the dispersion of a three-prism spectrograph is sufficient and it is found very convenient to be able to determine all alkali metals with a single spectrum plate.

Maxwell[75] could detect sodium as a main component of orthoclase with a 2 MW Q-switched laser.

Debras-Guédon and Liodec[27] found sodium (line 588·995 nm) as well as potassium (line 404·41 nm) in the microplasma produced by a ruby laser on the surfaces of nitrates and carbonates. Lithium is one of a number of elements that could not be determined without auxiliary electrodes between sample surface and objective.

### 4.2.9. Beryllium – magnesium – calcium – strontium – barium

Different from the elements of the first main group of the periodic system, the analysis of the alkaline earth metals, preferably also undertaken simultaneously for the whole group with a laser micro-analyser and a medium quartz prism spectrograph of the Q 24 type, does not present any difficulties.

All suitable analytical lines of beryllium lie in the ultra-violet. In checking regions of silicate spectra in micro-spectroscopy it is quite sufficient to consider only the lines Be I 234·86 nm and Be II 313·04 nm.

Y. Tuneko with an LMA 1 outfit could detect as little as $10^{-13}$ g Be in a standard silicate powder.[73]

The authors identified magnesium in silicates, arsenates, borates, oxides and in alloys by means of two very typical neighbouring groups of analytical lines, namely: the Mg II lines 279·08, 279·55, 279·81, 280·27 nm, the second and fourth of which always show strong self-absorption; and the although considerably weaker, yet very distinctive Mg I lines 277·67, 277·83, 277·98, 278·14, 278·3 nm.

The most important analytical lines of calcium in laser spectroscopy too are Ca II 396·85 and 393·37 nm; they do not exhibit self-absorption and can be recorded on medium quartz prism spectrographs in spite of the low dispersion near the upper wavelength limit of their working range.

Detecting strontium by using the main analytical lines Sr II 407·77 and 421·55 nm can also be carried out with lasers as accessories of quartz prism spectrographs. Debras-Guédon and Liodec[28] were able to find strontium in the microplasma produced by a laser on the surface of a carbonate, using the line Sr II 346·46 nm. In the spectrogram btained by them from a $BaSO_4$ sample the line Ba II 416·6 nm appeared amongst others.

After cross excitation in the auxiliary electrode gap between sample and microscope objective the authors also found on ORWO photographic plates WU-3 the principal Ba line 455·4 nm, although the plates are really not intended for application in this wavelength range; in addition to a number of weaker ultraviolet barium lines the Ba II lines 230·4 and 233·53 nm were furthermore present which are favoured as 'bypass lines' in conventional mineral and ore analysis. In the two-step procedure of laser micro-analysis these two last mentioned lines in the ultra-violet exhibit self-absorption to a considerable degree and are therefore not suitable for quantitative analysis.

Matsushita *et al.*[73] analysed with an LMA 1 outfit, producing craters of 100 μm diameter, Mg, Ca and Sr in silicate and carbon powder matrices hardened by phenolic resin and found for these elements detection limits of $10^{-13}$ g.

### 4.2.10. Boron – aluminium – gallium – indium – thallium

Boron and aluminium are also detected in the spark like discharge of the two-step method of laser micro-analysis with cross excitation between pointed carbon electrodes, use being made of pairs of arc lines. For work with the doublet B I 249·68 and 249·77 nm either a quartz prism or a grating spectrograph is required. Fig. 68 shows a section of the ultra-violet spectrum which the authors obtained with the first LMA outfit of Zeiss, including a Q 24 spectrograph, in 1964; this spectrum part is fully sufficient for the complete classification of a boron mineral sample. It can be seen that the line of the longer wavelength produces a considerably stronger blackening of the photographic plate than the line B I 249·68 nm, which is in good agreement with the spectrum tables in general use. The authors relied on these boron lines not only for the identification of different borates, but also for detecting boron in the lattice of various cyclosilicates; the boron content of rough tritomite samples found amongst the rocks of the Langesund Fjord, as well as that of very small needle-shaped tourmaline crystals, enclosed in quartz, could easily be detected; one firing of the neodymium glass laser model ZFL 750 radiating at 1060 nm with an energy of 0·2 J was sufficient. The spectrum was photographed with ORWO plates type WU-3.

In the routine analysis of aluminium the authors usually restricted themselves to the line pair Al I 394·40 and 396·15 nm in the visual spectrum region because no interference by other lines could be noticed; these lines can also be dealt with on a medium quartz spectrograph. Together with the two Ca II lines 393·37 and 396·85 nm on either side of them, these Al I lines represent useful reference lines for the measurement of weaker spectrum lines by interpolation.

Near the Al I line pair favoured by the authors lie two Ga I lines 403·3 and 417·2 nm which, although physically the most intense lines, so far could not be used in work with prism spectrographs of medium dispersion because of interference by cyanogen bands. In laser spectrochemical analysis there is no such coincidence of analytical lines and components of band systems. The authors were able to use the two Ga I lines in the rapid analysis of semiconductor materials, especially gold–gallium alloys. No disturbing self-absorption was observed. When interpreting germanite spectra obtained with the laser micro-analyser the authors found that of the two ultra-violet lines Ga I 287·4 and 294·36 nm, giving satisfactory results in ore and rock analyses, the second is distinctly stronger.

For determination of indium in special alloys the use of the In I lines

451·1 and 410·18 nm is recommended because the In I lines in the ultraviolet are diffuse on spectrum plates. An In II line at 244·79 nm can serve for checking. The authors could not see the 11·54 eV line In II 207·9 nm in InSb spectra taken on ORWO plates type WU-3.

The main analytical thallium line Tl I 535·05 nm cannot be utilized with an outfit consisting of the microscope model LM, fitted with an objective of long working distance, and a quartz prism spectrograph. In testing chemicals the authors relied on the line Tl I 377·57 nm and checked with the line Tl I 351·92 nm. There was no interference with the former line by bands even when carbon or graphite electrodes were used.

Maxwell in his qualitative analyses of silicate minerals with the laser microprobe could detect boron as trace element of plagioclase in anorthosite.[75] Apparently the determination of aluminium even with a ruby laser with Q-switch feedback is not always simple, as evidenced by the classification of aluminium as a trace element of plagioclase and as a minor constituent of potassium felspar.

Investigations carried out in France on aluminium alloys with use of the 7·66 eV line 305·7 nm proved that aluminium can be detected by imaging the laser microplasma on the spectrograph slit. Yet the presence of boron could evidently not be confirmed in work without cross excitation.[27]

### 4.2.11. Carbon – silicon – germanium – tin – lead

In any spectrum obtained with carbon electrodes on a Q 24 type spectrograph essentially the following lines appear: C I 247·8 nm, C II 250·9, 251·2 and 426·7 nm and C III 229·7 nm.

When carbon is to be detected, for instance in steel, suitable metal electrodes have to be chosen. At present the Zeiss laser micro-spectrum analyser can be linked with the quartz prism spectrograph model Q 24 or the plane grating spectrograph model PGS 2. This outfit is not always adequate for the determination of carbon in low and high alloy steels. The future development of a laser vacuum spectrograph is therefore desirable.

As only Si I lines are available for silicon determinations a minimum inductance setting of 30 $\mu$H on the mains supply unit for the auxiliary spark gap is indicated for the routine examination of silicates. For a quantitative silicon analysis about three laser pulses are required to obtain sufficiently strong lines. The most important Si I lines can be found fairly quickly if auxiliary electrodes of carbon are used and if the positions of the intense C I line 247·86 nm and of the C II lines 250·9, 251·17/251·2 nm are taken as reference. The Si lines in question are a group of six arc lines between 250·69 and 252·85 nm with the strongest at 251·6 nm. In addition a line of somewhat longer wavelength than this very typical group, namely the line Si I 288·16 nm, is commonly used.

In the spectrum of germanite the authors found the well-known Ge I lines in the ultra-violet. The strongest is that at 303·9 nm; the two Ge I lines at 265·16 and 265·12 nm are no longer resolved with slit widths of more than 15 μm. The line Ge I 326·95 nm is especially easy to find in the spectra of sulphide ores because the neighbouring lines Cu 324·75 and 327·396 nm on either side can be taken as reference.

The second strongest Sn I line 326·23 nm very conveniently also lies between these two main analytical copper lines; in a spectrogram of the Q 24 spectrograph enlarged twenty times it is about 13·5 mm from the physically strongest Sn I line 317·5 nm. The authors could not confirm this intensity relationship known from conventional spectroscopy in cassiterite spectra; to the eye the line 326·23 nm appeared to be somewhat stronger than the principal analytical line listed in spectrum line tables.

The main analytical lead line Pb I 405·78 nm can, in the cathode-layer technique, without short-time fractionation never be found with a prism spectrograph. According to an important finding, especially valuable for lead determination, this can, however, be done in laser micro-spectroscopy without any disturbing interference, even when carbon or graphite electrodes are used.

The characteristic line group Pb I 357·27, 367·15, 368·35 and 374 nm can also be recorded on a clear background and thus serve for lead detection in ores and alloys; so can the Pb I lines in the ultra-violet near the C II components 283·67 and 283·76 nm, namely the lines Pb I 280·2, 282·3, 283·3 and 287·3 nm. This series of lines, however, with arc-like discharges exhibits strong self-absorption and therefore only has limited value for quantitative investigations.

Different from silicon, lead can be detected with a laser microprobe with great sensitivity, even if the analysis is carried out without an inductance inserted in the circuit and although only arc lines are available for the determination of elements of the fourth group of the periodic system.

Experience has shown that focused red laser radiation will fairly easily excite silicon (in steel, wavelength 288·15 nm), tin (in metals, wavelength 326·23 nm) and lead (in metals, wavelength 357·27 nm) to a degree sufficient for the detection of the elements by directly concentrating the radiation from the plasma formed close to the sample surface on the slit of a powerful spectrograph.

Archbold, Harper and Hughes[7] could establish a transient emission of Ge IV and Sn IV lines by laser produced microplasma in work with a 1 MW Q-switched laser system.

American and Canadian scientists were able in analytical experiments to find lead traces in sphalerite and silicon traces in bornite, sphalerite, magnetite and galena. The determination of silicon as a main constituent of various types of plagioclase attempted in experiments at Newtonville sometimes failed because of the very small extent of action of a Q-switched solid-state laser.

### 4.2.12. Arsenic – antimony – bismuth

An appreciable number of special enrichment procedures have been evolved in the past for finding the volatile elements arsenic, antimony and bismuth in minerals, rocks and synthetic technical materials. The conventional spectroscopic determination of arsenic in particular has almost completely been abandoned in favour of photometric measurement of the concentration of molybdenum blue complexes.

When micro-analyses with laser excitation were carried out on sulphide, arsenides, arsenate and vanadate ores it emerged that the elements of the fifth main group of the periodic system can very easily be determined with a single laser pulse.

Generally only analytical lines in the ultra-violet can be considered in practical work. The authors could detect arsenic in small crystals of proustite, erythrite and cabrerite by means of the As I lines 228·8 and 234·98 nm; they found ruby and neodymium glass resonators equally suitable for this purpose. The As lines 286·04 and 289·87 nm are present in the spectra of skutterudite and arsenopyrite photographed with a 1·5 m Wadsworth grating spectrograph.

For identifying stibnite (antimonite) and polybasite and for confirmation the antimony lines Sb I 259·8 nm and Sb 252·85 nm were adequate.

Bi I 306·77 nm is the strongest line in sulphide and vanadate spectra; lower down in the spectrum follow the about equally useful bismuth arc lines 302·46, 298·9, 293·8 and 289·8 nm. In the authors' experience the line Bi I 280·96 nm was, even in typical bismuth ores, only weakly excited and can certainly not be considered to be a main analytical line. On the other hand the authors obtained a bismuth line of surprising strength at 286 nm.

Maxwell[75] could with the laser microprobe determine bismuth as trace element in galena.

Antimony and bismuth were not found in the first attempts at using as spectroscopic light source the microplasma produced by focused ruby laser radiation.

### 4.2.13. Phosphorus – sulphur – selenium – tellurium

Of the four elements phosphorus, sulphur, selenium and tellurium only the last two can be detected with the laser micro-analysers so far commercially available. In their examinations of semiconductor materials the authors used the two ultra-violet tellurium lines Te I 238·3 and 238·57 nm. The laser was fitted with a neodymium glass resonator and the sample was vaporized with a reflecting objective $\times 40/0{\cdot}5$, $\infty/2$. The photographic plates should be of the type of the ORWO plates ZU-2.

When high energy laser radiation is focused on selenium photocell elements it is, for instance, possible to detect in the different layers Se, Cd

Ag and constituent elements of the base plate such as Fe, Mn, Ni or Zn, and Cu.

For analysis of P and S a vacuum spectrograph is required or a very small electrode gap and extremely high capacitance (external ignition; analytical lines S III 349·7 nm, P III 323·4 nm).

### 4.2.14. Rhodium – palladium – platinum – osmium

Rhodium platinum alloys give spectra with very few lines; those available for determination of the two elements are Rh I 343·49 and 339·68 nm, Pt 265·9 nm and of about equal intensity 306·47 and 299·8 nm; between the last two lies a weaker platinum line Pt I 304 nm which is suitable for verification.

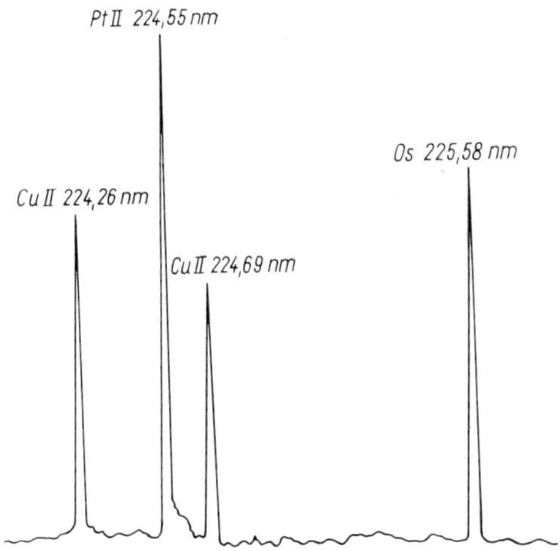


FIG. 89. Analysis of platinum minerals with the LMA 1 outfit.

Chemicals containing palladium give spectra showing the two neighbouring lines Pd I 340·46 and 342·1 nm, and a number of equally useful Pd II lines; one of these is isolated in a characteristic position at 285·46 nm and is often mentioned in the literature. A group of stronger Pd lines is crowded together within the wavelength region from 265·8 nm to 230 nm.

The determination of osmium in native platinum is based on the lines Os I 290·9 and 305·87 nm and can be checked with two other lines in the ultra-violet at 225·58 and 228·22 nm. Such verification is especially advisable when one works with slit widths of more than 15 μm. In the analysis of ore samples from Finland the authors found the very characteristic line groups Cu II 224·26/Pt II 224·55/Cu II 224·7/Os 225·58 nm (Fig. 89) and Os 228·26/Pt II 228·82 nm.

## 4.3. QUANTITATIVE ANALYSIS

At present only those research institutes which have at their disposal the necessary sets of solid, homogeneous and accurately analysed reference standards are in a position to undertake quantitative spot analyses of microgrammes of a sample with a laser micro-spectrum analyser. The stage of development reached can be characterized by the concept of a procedure dependent on the availability of key reference standards. The chemical nature of the standards must of course closely correspond to that of the samples under test. This problem should first be solved for metallurgical samples because under suitable conditions homogeneous standard melts can be produced of many alloys. The homogeneity may be tested by microscopy on a prepared polished section of the melt sample or by spectrochemical analysis. Small research institutes best procure metal and alloy standards of highest purity with exact analytical data from specialist firms. In the USA the Spex Industries Company offers various such standards and in England the firm of Johnson Matthey Metals Ltd of London.

A clear warning must be given not to practice quantitative analysis uncritically; Cerrai and Trucco[25] draw attention to the connection between the intensity of spectrum lines of a main alloy component and the structures of metal samples influenced by the working process or by recrystallization (matrix effects). When arc-welded Zircaloy samples were scanned with a sequence of laser pulses and also when uranium fuel rods were irradiated with a laser along a line on their surface, a remarkable intensity increase of Zr and U lines with decreasing grain size could be observed. Therefore quantitative evaluations of the local differences in the iron concentrations of alloys (0·07 to 0·10 per cent in the zirconium alloy and 250 g/t in uranium), unless appropriately corrected, are considerably in error. The Italian authors also noticed a more than two-fold intensity increase of the Al line 309·27 nm when, after analysing a completely recrystallized aluman alloy, they went on to a band grain structure produced by hardening. The intensity 'gain' for Mn, present in aluman in a 1·5 per cent concentration, amounted to a little over 14 per cent. Spectroscopic confirmation of these findings was obtained from a third type of sample which as a result of annealing had grain arrays differently recrystallized in its various parts, with the structure difference enhanced by stress.

Important for spectroscopic work is the conclusion that. by observation with polarizing optic closely resembling textures and structures of reference standard and test sample can fairly easily be recognized; and furthermore that the effects produced with a rather small non-switched laser of course must not be taken to be of a general nature.

Reference standards of non-conducting materials in the form of homogeneous solids are unfortunately not obtainable in the required quantities.

The following suggestions were made for the preparation of reference standards of substances that can be converted into solutions:

(a) Volatilization and fixing in an unexposed photographic emulsion; the silver contained in this gelatinous matrix material, also suitable for suspensions, serves as an internal standard and its concentration as basis for the calculation of the vaporized material quantity.[106]

(b) Freezing metallic salts out of aqueous solutions which gives homogeneous standards. These, however, have the disadvantage of not a very good possibility of detection.[60]

(c) Precipitation of sulphide mixtures through passing hydrogen sulphide into aqueous solutions. Sunahara *et al.* thereby obtained solid standards giving a better possibility of detection than method (b). Although the crater dimensions could be exactly measured these standards however were no longer homogeneous enough in small masses. Suitable working curves could be plotted for 0·01–0·5 per cent Ag and for 0·2–10 per cent Pb in CdS precipitates and for 0·3–8 per cent Fe and Cu in ZnS precipitates.[60]

Some users of electron beam microprobes, known since 1950, have experience in producing or obtaining appropriate reference standards. But it must be mentioned that quantitative analysis with the electron beam microprobe even today is still faced with a great many unsolved problems. Mallissa[71] wrote on this point in 1966:

'Die Durchführung quantitativer Analysen ist ein ziemlich komplexer Vorgang, wobei auf relativ viele Einzelfaktoren Rücksicht genommen werden muß, und es wird – besonders bei der Bestimmung mehrerer Elemente gleichzeitig – nicht möglich sein, die für jedes einzelne Element günstigsten Bedingungen anzuwenden, sondern es wird fast immer zu einem Kompromiß kommen (p. 124 of ref. 71).' [A quantitative analysis is a rather complicated operation in which many single factors have to be taken into account, and it may not be possible, especially in the simultaneous determination of several elements, to adopt the conditions that are the best for each individual element; almost in every case a compromise will have to be made.]

In quantitative laser micro-spectroscopy also various points must be considered. A specially favourable feature is the absence of any background on the spectrum plate in the majority of cases. The usual corrections therefore need not be applied, mainly then when the inductance of the mains supply unit of the LMA 1 outfit is set to 30 or 60 $\mu$H.

Different from excitation by conventional spark sources and by d.c. or a.c. continuous arcs, self-absorption very frequently occurs especially with laser excitation[32] and sometimes also in the two-step procedure with the LMA 1 model. As a result well-known analytical lines can, even at relatively low concentrations, only be used in semi-quantitative work with curved working graphs, or cannot be used at all.

Often only a smaller concentration range can be evaluated with one particular analytical line than in established macroscale procedures. Usually, however, a sufficient number of spectrum lines of the element to be determined quantitatively is available to cover a wider concentration range by means of several sectional working curves, a fact considered a 'drawback' of optical as compared with X-ray spectra.

Different from the conditions in qualitative work, the operating parameters of the laser and the auxiliary electrode gap have to be specially well controlled for quantitative analyses. It has been indicated in the literature that selected small sample parts can be vaporized with particularly good repeatability by using a Q-switched resonator and that consequently the line intensity can be better reproduced[103]. This statement cannot be generally confirmed without qualification because the reproducibility of sample vaporization with a Q-switched laser also depends on exactly maintaining many operating characteristics of the laser emission (pulse

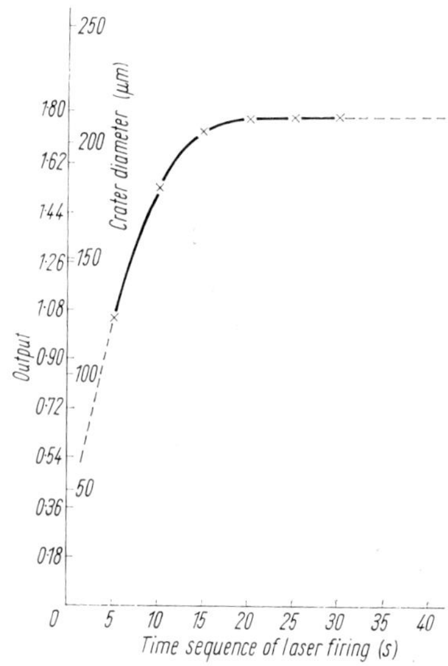


FIG. 90. Change of crater diameter in steel with repeated laser firings with LMA 1 laser analyser (neodymium glass resonator).

repetition, output) through controlling the flash tube voltage, resonator temperature, concentration of switch liquid or angular velocity of rotating reflectors, overall instrument adjustment etc. According to Sunahara[60] crater diameters can with a Q-switched pulsed laser (Jarrell-Ash model) be better reproduced in stainless steel and in photographic emulsions (deviations of 2·6 and 3·6 per cent) than in brass and in Cu–Fe sulphides. The reproducibility was especially bad for aluminium and ceramic samples (dispersion of 16·7 and 25·3 per cent).

In the literature the reproducibility of the output from actively switched ruby lasers, of the craters produced by them and of the spectra obtained in the two-step (cross excitation) procedure has been much discussed.[60,105,121] Although the experimental results cannot be assumed without reservation to be valid for four-level lasers or non-switched systems they are of interest in various respects. The authors of the cited publications unanimously stress the difficulty of adequate cooling of the ruby lasers. Snetsinger and Keil[121] found that the standard deviations of the Q-switched ruby laser output of $\pm 1{\cdot}8$ per cent (pulse repetition rate 1 $min^{-1}$) and $\pm 1{\cdot}2$ per cent (pulse repetition rate 0·1 $min^{-1}$) were remarkably better than those of the crater volumes. As possible explanation interactions between laser radiation and microplasma, and time variations of the spike separation are mentioned.

Adequate cooling of the resonator head is of very great importance. Furthermore, one has to remember that the various resonator materials often have very different thermal conductivities; hence the necessity of accurately controlling the cooling periods between consecutive laser firings (Fig. 90).

The electrode distance likewise must always be exactly set and maintained; therefore provision must be made in a laser micro-spectrum analyser for centring and aligning the auxiliary electrodes by microscope. It is found that in the various procedures of quantitative laser microspectrochemical analysis very often specially steep working curves have to be used (Fig. 91).

Three possible working modes theoretically exist also in quantitative work, namely those of area analysis, line analysis and point analysis:

*1. Area analysis.* This type of analysis can often be carried out by conventional spectrochemical methods on suitably prepared metal samples as average determination of elements uniformly distributed over an area. The use of a pulsed solid-state laser offers advantages for the direct nondestructive analysis of non-conducting materials. Mechanical size reduction or dissolving of the sample are not required. The minute sample part selected for irradiation is vaporized by the focused laser radiation. This problem of quantitative analysis was tackled at the Institute for Spectrochemistry and Applied Spectroscopy, Dortmund, in 1965.[33] The sample was scanned in two directions at right angles by means of an automatic drive with step motors, as illustrated by the drawing and diagram of Figs. 92 and 93. Sixty-five to a hundred laser pulses were recorded in one spectrum. The outfit had a Q-switched ruby resonator of 70 mm length and 6 mm diameter, 0·1 J output, repetition rate of 12·5 Hz, with water cooling for flash tube and resonator; it worked without cross excitation. The method is one of macro-analysis in which the required quantity of material is covered in a series of point analyses.

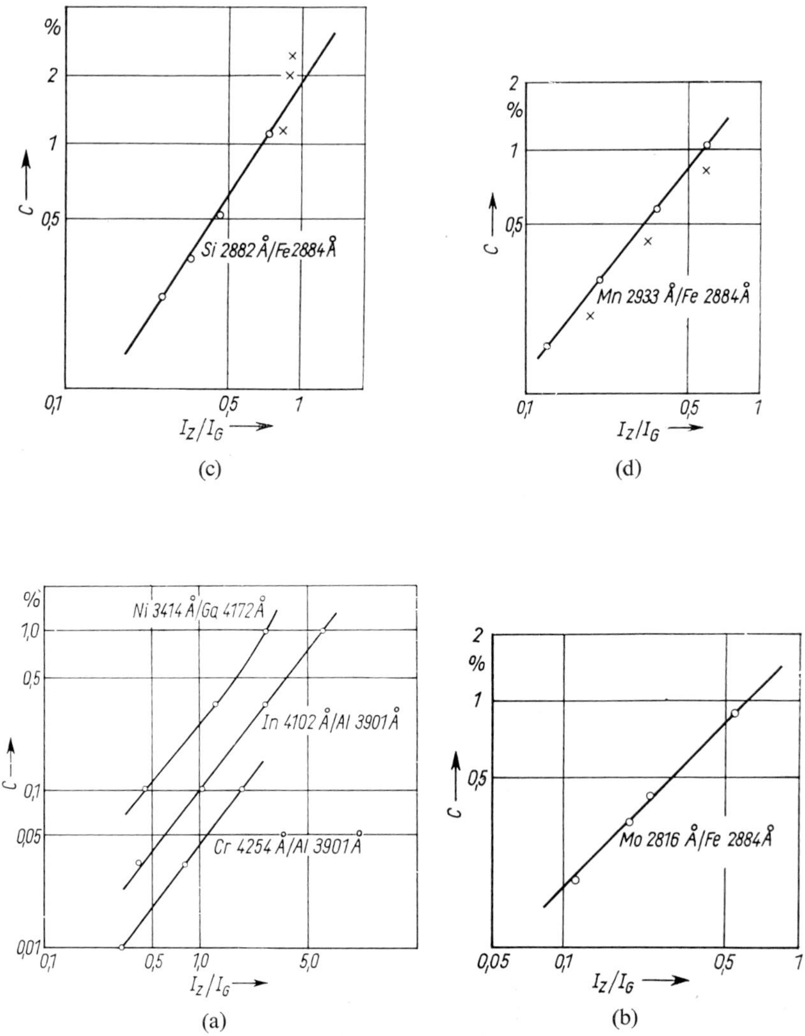


FIG. 91. Working curves for area analysis with a series of laser pulses according to Felske, Hagenah and Laqua.[33]

(*a*) For $Al_2O_3$.

(*b*) to (*d*) For steel (the values near the Si and Mn curves indicated by crosses were obtained for cast iron).

2. *Line analysis.* A line analysis for quantitative purposes can in many cases be carried out more efficiently with an electron beam microprobe if the scanning range is 0·5 mm or a little less. There are, however, test

objects that cannot be scanned with an electron beam, as for instance varnish samples. The advantages offered by the LMA outfit are the possibility of analysing non-conducting materials without the necessity of applying a conducting coating, of determining elements with very low atomic numbers (mainly lithium and beryllium), of analysing uneven and large samples and of carrying out analyses over greater distances. In the line analysis carried out experimentally by the authors at right angles to the boundary layer of two metals, to a soldered joint and to a welding seam, the line on the sample had to be scanned point by point and the corresponding number of spectra had to be measured on a microphotometer.

*3. Point analysis.* In point analysis a very small zone of the sample has to be investigated because of its chemical nature which differs from that of its surroundings.

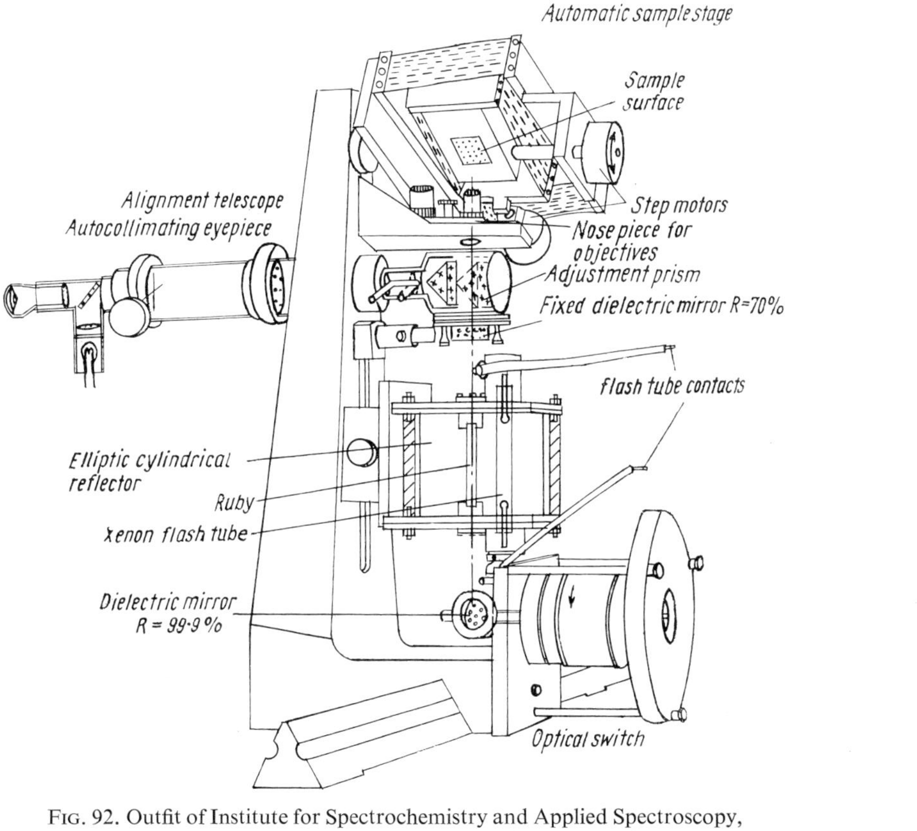


FIG. 92. Outfit of Institute for Spectrochemistry and Applied Spectroscopy, Dortmund (G.F.R.), for quantitative area analysis with laser excitation according to Felsk *et al.*[33]

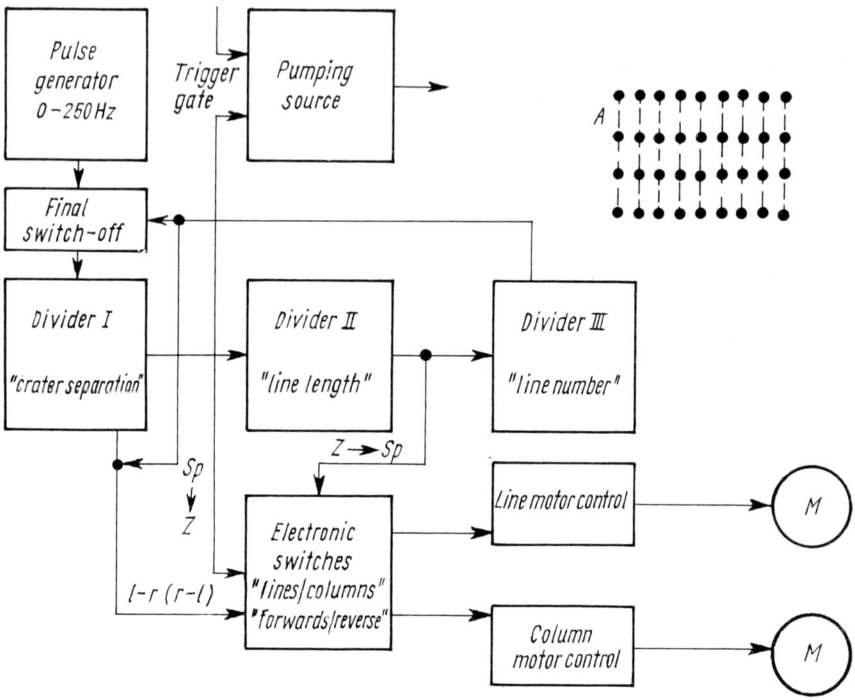


FIG. 93. Circuit block diagram for controlled sample movement in area analysis according to Felske *et al.*[33]
A: Trace of focal spot scanning the sample surface.

For testing the reproducibility of such a determination a chromium steel 145 Cr V 6 was analysed with the LMA 1 laser micro-spectrum analyser and the quartz spectrograph Q 24. The sample had the following element concentrations: Ca 1·47, Si 0·29, Mn 0·46, Cu 0·16, Ni 0·09, V 0·08, Cr 1·40 per cent. The work was carried out with a slit width of 20 μm a neodymium glass resonator, refracting a objective × 16/0·2, carbon electrodes, capacitance of flash tube a capacitor bank 508 μF, capacitance of spark gap 2 μF and its inductance 60 μH. The craters had diameters of 230 μm. Ten spectra, each produced by one laser firing, were taken on each of four photographic plates (ORWO WU 3, blue sensitive, extra hard) with the use of two alternative mains supply units for the LMA 1 outfit. The relative standard deviations of the density values of the chromium line Cr I 425·43 nm were:

| | |
|---|---|
| Plate 1, 1st supply unit | 1·4 per cent |
| Plate 2, 1st supply unit | 2·0 per cent |
| Plate 3, 2nd supply unit | 2·5 per cent |
| Plate 4, 2nd supply unit | 2·3 per cent |

The variance of the density values of the recorded spectra (Plates 1 to 4) was found to be 2·5 per cent.

In all these different forms of quantitative laser micro-spectrochemical analysis the particular advantages of the separation of vaporization and excitation become apparent. They are:

(a) No loss at the striking of the arc
(b) No wandering of the arc
(c) No 'locking' of the arc or spark discharge
(d) Uniform single discharge
(e) No discharge forms without vaporization
(f) Sample cannot be ejected from the bore hole in the electrode or form fused beads
(g) No interfering fractional distillation
(h) Emission of CN bands strongly suppressed

Ryan *et al.*[114] carried out quantitative examinations of the $Al_2O_3$ content of a glass knot with an American laser probe; whereas 10 per cent and 17 per cent of the oxide were found in the centre of the inclusion, the glass matrix only contained 2·6 per cent $Al_2O_3$. They also tested a chromium spinel and its neighbourhood in a glass sample with the following result:

TABLE 20. Quantitative analysis of a chromium spinel inclusion in glass according to Ryan et al.[114]

| *Laser pulse* | *Sample part* | $Cr_2O_3$ (per cent) | $Al_2O_3$ (per cent) | $Fe_2O_3$ (per cent) | $MgO$ (per cent) |
|---|---|---|---|---|---|
| 1 | Spinel core | 35 | 30 | 21 | 12 |
| 2 | Green shell | 30 | 11 | 30 | 25 |
| 3 | Adjacent glass | 10 | 4 | 5 | 18 |
| 4 | Host glass | <0·1 | 2·2 | <0·5 | <0·5 |

From these quantitative results it was inferred that Fe and Mg are more soluble in the tested glass than Cr. Unfortunately no information is given on the reference standards used in the work.

The authors found that powder reference standards, when very well pulverized (grain size $\leq 2$ μm) and pressed into solid disks, are suitable for preparing working curves; but the operating conditions of the instrument must be controlled so that the craters produced on the pressed disks have diameters of more than 100 μm. Such standards are mainly the well known granite, syenite and diabase samples G 1, S 1 and W 1 or synthetic oxide mixtures, obtainable from Johnson Matthey, London (Spec-pure grade).

Especially good homogeneity can be achieved in the following way. The mentioned powder reference standards are digested in a platinum crucible with sodium carbonate or borax, the resulting material is pulverized and then pressed into small disks. The disadvantage of the method is the lowering of the concentration through the digesting agent and the transformation of the silicates into carbonates or borates so that chiefly reference standards for the analysis of carbonates and borates result. The pressed disks can only under certain conditions serve for a silicate analysis. Possibly existing matrix effects still have to be investigated more thoroughly.

According to Whitehead and Heady[132] trace elements dispersed in homogeneous matrices could be determined with standard deviations of about 10 per cent. Matrix effects in the case of vaporization by laser radiation are in general restricted to the relationship between the preparatory treatment or crystallization history of the sample and the affected material quantity; the effects are due less to the influence of other elements than to the grain size of the pressed powder samples. An investigation carried out in Milan on $UO_2$ particles,[25] for instance, has shown that grains that have been less strongly implanted in the test substance by sintering processes lead to a greater possibility of detection sensitivity than those in more densely packed matrices.

A third possibility of obtaining homogeneous solid targets for quantitative micro-analysis from standard reference powders is the manufacture of special glass types; these however can only be used with commercial micro-spectrum analysers normally fitted with non-switched solid-state pulsed lasers if they are intensely coloured and are particularly absorbing at 1060 or 694·3 nm.

Karyakin, Kaigorodov and Achmanova[58] occasionally formed pressed disks from silicate and standard reference powders mixed with potassium bromide. They found that a neodymium glass laser with output energy of 10 to 12 J can be satisfactorily employed for the quantitative analysis of geological samples in quantities of milligrammes. Results obtained for Si, Al, Fe, Mn in wet chemical analysis and in the two-step procedure agreed within satisfactory limits:

| | *Chemical Analysis* | *Two-Step Process (Vaporization by Laser)* |
|---|---|---|
| | per cent | per cent |
| Si | 16·6 | 14·9 |
| Al | 4·9 | 4·6 |
| Fe | 22·7 | 26·2 |
| Mn | 0·18 | 0·189 |

A quantitative method for determining 0·02 to 2 per cent Cr and 0·05 to 1 per cent N: in steel samples has been developed in Paris. The evaluation of the test results gave for a chromium concentration of 0·92 per cent a standard deviation of 0·04. In this emission macro-procedure with a laser of 100 J output energy crossexcitation was not needed.[129b]

The authors performed the following experiment. Reference standards in disk form were prepared for the quantitative analysis of several constituents of carbonate inclusions. As natural rock standards, namely Granite G-1, Syenite S-1 and Diabase W-1 from the U.S. Geological Survey were available in powder form from previous work, batches of small quantities of these materials were digested in a platinum crucible with a four-to five-fold quantity of a mixture of sodium carbonate and potassium carbonate. The solidified melt was pulverized and pressed into disks. The LMA 1 micro-spectrum analyser was fitted with a neodymium glass resonator, the flat-field achromat $\times 16/0 \cdot 20$ and with carbon electrodes. The flash tube voltage was 2 kV, the capacitance of its capacitor bank 508 μF, the excitation voltage 2 kV, the capacitance of the spark gap 2·5 μF and its inductance 60 μH. The slit of the spectrograph model Q 24 was set to a width of 14 μm. A three-step filter was used and ten spectra were photographed on each of several plates. Care was taken to

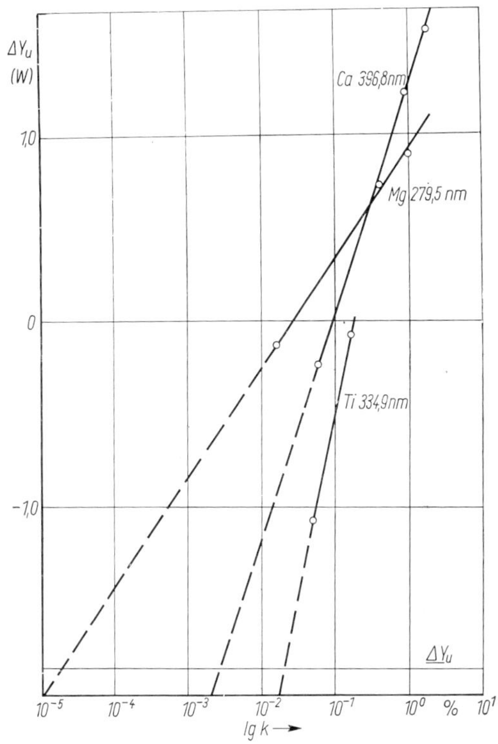


FIG. 94. Working curves for determining magnesium, titanium and calcium in carbonates with an LMA 1 outfit.

maintain identical working conditions for each analysis; the electrode distance was re-set after each exposure.

Three working curves for the determination of the elements Mg, Ti and Ca in carbonates are shown in Fig. 94. The following lines were utilized with the technical data taken from the tables of Saidel, Prokofjew and Raiski (Berlin 1961):

TABLE 21. The spectrum lines utilized for the working curves of Fig. 94 and the determined limits of detection

| | *Wavelength* (nm) | *Intensity* *Arc* | *Spark* | $\Delta Y_U$ (per cent) |
|---|---|---|---|---|
| Mg II | 279·5 | 150 | 300 | $2 \times 10^{-5}$ |
| Ti II | 334·9 | 125 | 800 | $2 \times 10^{-2}$ |
| Ca II | 396·8 | 500 | 500 | $3 \times 10^{-3}$ |
| (Ca II | 393·3 | 600 | 600 | $1 \times 10^{-3}$) |

The relative limit of detection, according to H. Kaiser, '... ist dadurch bestimmt, daß sich das analytische 'Signal', der Meßwert der Linienintensität, nicht mehr sicher aus den statistischen Schwankungen der Meßwerte für die Untergrundintensität, dem Störpegel, heraushebt.' (*Optik*, 1964, **21**, 311.) [is reached when the analytical 'signal', the measured value of the line intensity, is no longer clearly distinct from the statistical fluctuation of the measured values of the background, that is the noise level.]

Kaiser has defined the relative limit of detection numerically: 'An der Nachweisgrenze soll der Schwärzungsunterschied, mit dem die Linie sich über dem Untergrund erhebt, das $3\sqrt{2}$ fache der mittleren Schwärzungsschwankung sein.' (*Spectrochim. Acta*, 1947, **3**, 65.) [At the limit of detection the difference between the densities of line and background should be $3\sqrt{2}$ times the mean fluctuation of the density.]

Laser micro-spectrochemical analysis has the great advantage of extremely weak background radiation. Using ORWO plates type WU 3 (blue sensitive, extra hard) the authors could unmistakably establish the presence of a line of density 0·006, that is at a Seidel transformation value $\Delta Y_U$ of $-1{\cdot}86$. The concentration at the limit of detection was determined from the graph. For magnesium (line 279·5 nm) the relative limit of detection was found to be $2 \times 10^{-5}$ per cent. For a vaporized quantity of material of $10^{-6}$ g this corresponds to an absolute limit of detection of the order of $10^{-13}$ g (Table 22).

An analysis may be undertaken without incorporating the standard reference ('key element') spectrum on each plate only if identical quantities of test sample and standard reference material are vaporized under the same excitation conditions. If then the produced crater diameters are not the same, the flash tube voltage can be altered to reduce or increase

TABLE 22. Relative and absolute limits of detection in laser micro-spectrochemical analysis

(*a*) Cu, Al and Fe alloys[121]
(*b*) Pressed disk of graphite with reference standard powder[121]
(*c*) Pressed disk of standard amphibolite powder[93]
(*d*) Pure zinc[104]
(*e*) Pressed disk of $Al_2O_3$ with reference standard powder[132]
(*f*) Steel[131]
(*g*) Pressed disk of reference standard powder[73]
(*h*) Alloys[60]

*a, b, d, e, h*: analysed with the Jarrell–Ash laser microprobe 45–601, ruby rod, Q-switch

*c, f, g*: analysed with the Zeiss laser micro-spectrum analyser LMA 1, neodymium glass rod

| | *Relative limits* (ppm) | | | | | *Absolute limits* (g) | | |
|---|---|---|---|---|---|---|---|---|
| | (*a*) | (*b*) | (*c*) | (*d*) | (*e*) | (*f*) | (*g*) | (*h*) |
| Ag | 10 | < 1000 | 10 | | | | $10^{-13}$ | |
| Al | 200 | < 1000 | | | | $2{\cdot}5\times10^{-9}$ | | |
| As | | > 1000 | | | | | | |
| B | > 32 | < 1000 | | | 11 | | | |
| Ba | | < 1000 | | | | | | |
| Be | 12 | < 1000 | 10 | | 0·8 | | $10^{-13}$ | |
| Bi | 180 | < 1000 | | | | | | |
| Ca | | < 1000 | | | 24 | | $10^{-13}$ | |
| Cd | | < 1000 | | | | | | |
| Ce | | < 1000 | | | | | | |
| Co | > 700 | < 1000 | 100 | | | $1\times10^{-9}$ | | |
| Cr | 60 | < 1000 | 100 | | 11 | $1{\cdot}5\times10^{-10}$ | | $2{\cdot}5\times10^{-10}$ |
| Cu | | < 1000 | 10 | 10 | | $5\times10^{-11}$ | $10^{-13}$ | $5\times10^{-10}$ |
| Fe | 240 | < 1000 | | 50 | 45 | | | |
| Ga | 80 | < 1000 | | | | | | |
| Ge | | < 1000 | 10 | | | | | |
| Hf | | > 1000 | | | | | | |
| Hg | | < 1000 | | | | | | |
| In | | < 1000 | | | | | | |
| Li | | < 1000 | | | | | | |
| Mg | | < 1000 | | 10 | | | $10^{-13}$ | $3\times10^{-10}$ |
| Mn | 40 | < 1000 | 100 | 20 | | $6\times10^{-11}$ | $10^{-13}$ | $5\times10^{-10}$ |
| Mo | | < 1000 | 100 | | 21 | $4\times10^{-9}$ | | |
| Na | | < 1000 | | | | | | |
| Nb | | < 1000 | | | | | | |
| Ni | 210 | < 1000 | 100 | | 25 | | | $1\times10^{-9}$ |
| P | > 1100 | < 1000 | | | | | | |
| Pb | > 200 | < 1000 | 100 | | 30 | | | $4{\cdot}5\times10^{-10}$ |
| Sb | > 500 | < 1000 | | | | | | |
| Sc | | | 100 | | | | | |
| Si | 270 | < 1000 | | 20 | | $3\times10^{-10}$ | | $1\times10^{-8}$ |

Table 22 cont.

| | *Relative limits* (ppm) | | | | | *Absolute limits* (g) | | |
|---|---|---|---|---|---|---|---|---|
| | *(a)* | *(b)* | *(c)* | *(d)* | *(e)* | *(f)* | *(g)* | *(h)* |
| Sn | 480 | < 1000 | 10 | | 50 | | | $3\times10^{-9}$ |
| Sr | | < 1000 | 100 | | 10 | | $10^{-13}$ | |
| Ta | | > 1000 | | | | | | |
| Te | > 400 | < 1000 | | | | | | |
| Th | | < 1000 | | | | | | |
| Ti | 280 | < 1000 | 100 | | 20 | $5\times10^{-10}$ | | $4\times10^{-10}$ |
| Tl | | < 1000 | | | | | | |
| U | | < 1000 | | | | | | |
| V | 830 | < 1000 | 100 | | | $7{\cdot}5\times10^{-10}$ | | |
| W | > 480 | < 1000 | | | 400 | | | |
| Y | | | 1000 | | | | | |
| Zn | | > 1000 | | | | | | $5{\cdot}5\times10^{-9}$ |
| Zr | | < 1000 | 1000 | | | | | |

the diameter as required. When working without a 'key element' spectrum it is advisble to use the intensity ratio of two lines, namely the line of the element to be determined and a suitable line of the basic element which is always present in the reference standard and test sample in the same amount. This condition again is easier to fulfil for metallic than for non-metallic samples. A suitable element can, if necessary, be added to the reference standard, but not to a test sample of microscopic dimensions embedded in a matrix. The authors, therefore, tried to use carbon lines, for example C 247·8 nm, as homologous lines. Evidently the conditions of volatilization are not the same for the electrode carbon and the test sample. It still has to be carefully investigated how far these conditions are comparable. The excitation voltage for the electrodes, the capacitance and the inductance can be set very accurately on the mains supply unit. The electrode gap has to be checked after each exposure for quantitative work; it is adjusted with the help of the eyepiece graticule used with the flat-field achromat ×4/0·5. The authors experimented with different gap settings and compared the intensities of the carbon line C 247·8 nm and the silicon line Si 288·1 nm in ten spectra. Fig. 95 shows that the intensity values of the two lines follow a similar curve. The standard deviation of the intensity ratios $\Delta D$ Si/C of the ten spectra was 6 per cent (see working curve Fig. 96). The fluctuation of the transformed density value of the silicon line alone was 21 per cent.

If in the analysis of an inclusion in steel the average composition of the surrounding basic material has to be known for comparison purposes, the homogeneity of this material must be examined by microscope. The authors had at their disposal a set of steel samples that had been very accurately analysed chemically and for which they were indebted to the VEB Edelstahlwerk '8. Mai 1945'. The samples were plain carbon, chromium

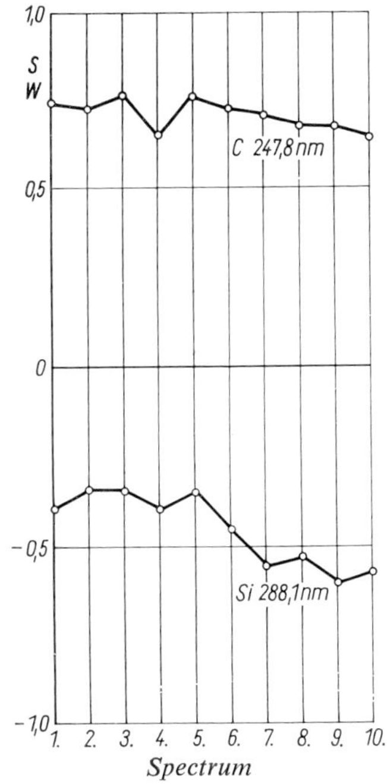


FIG. 95. Intensity variations of carbon and silicon lines at different electrode separations.

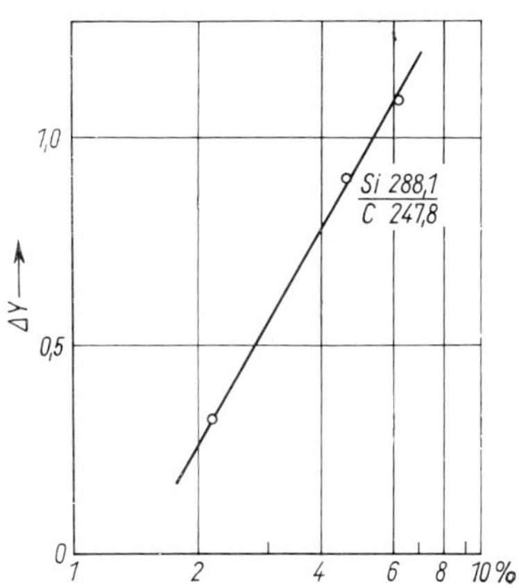


FIG. 96. Working curves for determining silicon in steel with the LMA 1 outfit.

and high-alloy chromium, nickel-chromium and nickel-chromium-molybdenum steels. The following concentration ranges were covered with them.

TABLE 23. Steel components and their analytical lines

| *Element* | *Concentration* (per cent) | *Analytical line* (nm) | *Intensity* *Arc* | *Spark* | *Electrodes* |
|---|---|---|---|---|---|
| C | 0·11 to 1·47 | 247·8 | 400 | – | Mg |
| Si | 0·21 to 0·80 | 288·1 | 500 | 400 | Mg or C |
| Mn | 0·25 to 0·69 | 257·6 | 300 | 2000 | Mg or C |
| Cr | 0·46 to 13·4 | 425·4 | 5000 | 1000 | Mg or C |
| Cu | 0·10 to 0·30 | 324·7 | 5000 | 2000 | Mg or C |

The determination of molybdenum, nickel and vanadium in concentrations of the order of 0·01 per cent and less is difficult because the analytical lines within the wavelength range of a spectrograph of the type of model Q 24 are of low intensity or suffer from interference, especially by the lines of the very complex iron spectrum.

The carbon content can only be determined by means of pure metal electrodes. The authors used pointed magnesium rods. The tips of metal electrodes, however, burn away much quicker than those of carbon electrodes so that metal electrodes have to be replaced after each analysis.

Phosphorus and sulphur cannot be determined with the standard outfit LMA 1 which does not include a vacuum sample chamber and a suitable vacuum spectrograph.

Chromium in steel can very satisfactorily be evaluated with a standard deviation of $\pm 2{\cdot}5$ per cent by means of the working curve of Fig. 97. A useful homologous line pair is Cr 425·4 nm and Fe 426·0 nm.

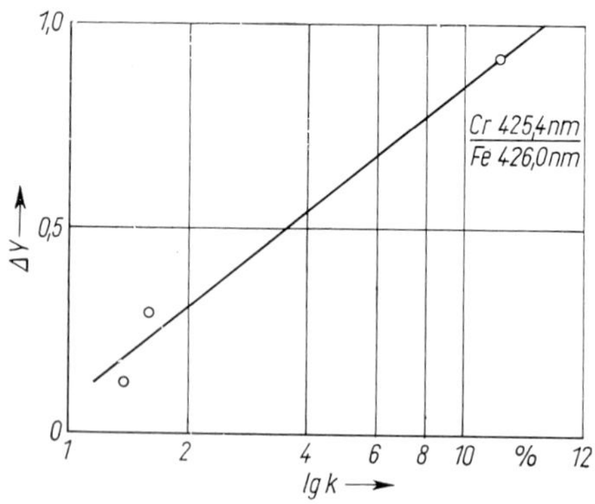


FIG. 97. Working curve for determining chromium in steel.

Blackburn *et al.*[15] polished sections of slabs of garnet containing gneisses, about 10 mm thick, and quantitatively analysed the Fe, Mg, Mn and Ca content of the nesosilicates in the rock matrix. Taking into account the interrelationships of crystal chemistry they laid down working curves for intensity ratios of the type

$$\frac{I_{\mathrm{Fe}\,274}}{I_{\mathrm{Fe}\,274} + I_{\mathrm{Mg}\,285\cdot2} + I_{\mathrm{Mn}\,403}}$$

A very good insight into the chemical nature of mixed crystals is provided.

The reference standards were silicate glasses, fused at +1400 °C in aluminium moulds, and the standard rock Diabase W-1, modified in a similar way. The use of a computer IBM/360 Fortran IV is worth noting. Out of a total of 222 analysed garnet crystals less than 10 per cent gave unsatisfactory results.

## 4.4. DEVELOPMENT TRENDS

Five years after invention of the solid-state laser and three years after the practice of laser micro-spectrochemical analysis had been suggested, outfits were commercially available that definitely have to be designated as accessories for standard spectrographs. In the future no doubt a large research instrument for laser spectroscopy will be developed which will provide a very much wider choice of microscopy procedures, will accommodate several interchangeable laser heads and will allow extension of the mains supply unit. Also the reliability of performance will be improved.

The routine analysis of larger sample sections and similar special problems in spectroscopy call for greater speed of operation with the same energy of the individual laser pulses. Here the experience can be turned to account that was gained in producing micro-processing and welding apparatus. Various models of such instruments are already known.

The Maser Optics Inc., Boston, Mass. makes a small Low Cost Metal Working Equipment MW-100 with a mains supply unit of 800 J output energy. The flash tube voltage is up to 2 kV, the capacitance of the flash tube capacitors 400 μF. The water-cooled ruby resonator of 6·35 mm diameter and 76·2 mm length emits 0·5 J (Model MW-101: 0·8 J) with a pulse repetition rate of over 1 $s^{-1}$. Sample details are viewed with a magnification of ×72.

The 604 Microwelder of TRG has a pulse energy four times as great, that is 2 J. The diameter of the focal spot at a working distance of 5 mm is about 50 μm.

The Microwelder Model 500 of the Hughes Aircraft Co. has an energy output of between 0·1 and 2 J. The diameters of the welding spots are between 127 and 500 μm. The pulse output energy is 1 J at a pulse repetition rate of 12 $min^{-1}$.

The Linde Laser Welder of Union Carbide, Linde Division, N.Y., works fully automatically and delivers 12 pulses per minute, each of 10 J; the welding spots have a diameter of about 380 μm.

The Korad Corporation, Santa Monica, developed a laser welder giving 5 to 15 J with a repetition rate of 1200 pulses per hour (*Electronic News*, 1964, **9**, 434).

The model RL 40 U of the C.G.E., Marcoussis, delivers an output energy of 4 J once per minute; it is fitted with objectives of 15, 30 and 45 mm local lengths.

At the Leipzig Fair of 1965 the USSR showed the welding equipment SU-1 with an optical quantum generator having, without objective, a radiation energy level of not less than 0·7 J. The outfit runs on a mains supply of 220 V, 50 Hz and takes 1·2 kW. With forced air cooling at 1 to 2 atm, four laser pulses per minute are possible with a pulse duration of 5 ms. Wires of about 0·1 mm diameter can be welded with the instrument; joints were successfully made of the following pairs of materials: Au–Si, Ta–Cu, Au–Ge, Ni–Al and Au–invar steel.

In spite of the relatively high repetition rates such outfits are normally not suitable for micro-spectrochemical analysis because focusing lenses very close to the work piece are used instead of microscope objectives with long working distances. Only the high repetition rate laser could therefore be adopted in spectroscopy. This of course entails some expense, but nevertheless these ruby lasers with high repetition rate will in future probably be the right component of the routine as well as of the research outfit, of the former especially when designed as an automatic model.

For the field work of geologists and mineralogists a portable instrument is required. Standard solid-state lasers need a supply unit that is much too heavy, and often the laser head itself is too large. Therefore efficient solid-state lasers of very small dimensions should be specially considered for the preliminary design concepts of a light and portable instrument.

What is wanted is a hand bag-size ruby laser run off a battery. Gürs and Westermeier of Siemens and Halske in Munich[47] designed what might be the forerunner of such a development, namely a laser unit with a resonator housing of dimensions 75 × 100 × 155 mm. The pump energy is about 1·6 to 2·4 times the threshold value (63 J) which of course is inadequate for spectroscopy; commercial photoflash lamps could be incorporated as generators.

Characteristic for all development trends in instrumentation is the very great speed of advance (see Table 24). Examples are the straight xenon flash tubes which superseded the helical forms, extension of the available range of resonators, amongst others by use of garnets, reduction in the size of mains supply units, more efficient cooling systems, replacement of mechanical and electrical modulating systems by Q-switching with phototropic substances.

There have been indications since 1966 that in future neodymium glass will generally take the place of ruby in the lasers of standard instruments. Following the VEB Carl Zeiss, Jena, Jarrell–Ash and LOMO have also changed their basic outfits for work at the wavelength 1060 nm.[53]

There is no doubt that for preliminary micro-spectrochemical investigations in the days to come the following combination of instruments will preferably still be chosen: solid-state laser (output energy 0·5 to 1 J, non-switched, with optional Q-switch attachment), microscope on optical bench, photographic spectrograph.

TABLE 24. Rate of development of microprobes

| | *Electron probe* | | *Laser probe* |
|---|---|---|---|
| 1927 | Busch establishes discipline of electron optics at Jena | 1960 | Maiman constructs the first solid-state laser |
| 1949 | Castaing describes the first X-ray micro-analyser with electron beam excitation | 1962 | Brech and Schuch report on the first laser microprobe with a Q-switched ruby laser |
| 1956 | Cosslett suggests the scanning electron probe | 1964 | VEB Carl Zeiss, Jena, develops the first commercial outfit for laser micro-spectroscopy model LMA with a non-switched solid-state laser |
| 1960 | Fourteen electron probes are working, about fifteen being constructed | 1967 | LOMO of Leningrad develops the MSL-2 model |
| 1964 | Two hundred instruments of twelve different makes are working | 1968 | Jeol, Tokyo, and Shimadzu, Kyoto, develop prototypes |
| 1966 | Six hundred instruments are working | 1973 | VEB Carl Zeiss, Jena, develops LMA 10[135] |
| 1969 | About eight hundred instruments are working | | |
| | Time interval between start of development and availability of several commercial instruments | | |
| | 30 years | | 3 years |

Among the problems dealt with in recent work are:

1. Improvement in the detection sensitivity by a factor of 20 to 50 through condensation of the material vaporized by laser radiation in a graphite cell which is used like a conventional arc electrode in atomic spectrum analysis.[9]

2. Prolonging the emission period of the plasma and increasing the signal to background intensity ratio by replacing the auxiliary electrodes

with a high-frequency oscillator. Typical features of experimental units for such an electrodeless high-frequency discharge are a frequency of 80 MHz and a maximum pulse output of about 1 kW for 200 ms. The normal emission period of 4 μs of a magnesium plasma could be extended by a factor of $4 \times 10^4$.[78]

3. Introduction of an infra-red transmitting collecting plate between focusing optic and sample; the thin films condensed from the microplasma make it possible to detect valence and deformation vibrations by means of an infra-red spectrometer. Karyakin *et al.*[1] in this way systematically compared with an UR 10 spectrophotometer silicate and carbonate minrals, vitreous rocks and magmatic samples with their condensation products. Theoretically further analytical evaluation of such condensates by other than spectroscopic methods is also possible.

4. Safety measures for the vaporization of radioactive materials.[89,126]

Special instrumentation is required for particular tasks of quantitative spectrum analysis in the microgramme range. For the principal application in this field, namely steel analysis, such instrumentation consists of a Q-switched solid-state laser and a vacuum direct-reading spectrograph with photomultipliers as detectors, linked by a microscope for work with reflected light. There are also many indications that, in future in spectrochemical analysis on the macroscale, solid-state lasers will be used more and more for vaporization. High-energy lasers with pulse energies of up to about 100 J will then probably be chosen, together with multi-purpose analytical outfits for alternative photographic and photoelectric recording of the spectra, and finally automatic laser analysers for process control in the metal producing and metal working industries.

Two trends are now noticeable in the development of Q-switched systems.

Since 1967 histochemistry has asked for laser excitation of picogramme quantities of material for quantitative cation analysis of microtome tissue sections. So far detection of six elements has been found to be feasible with craters of 10 μm diameter and 1 to 3 μm depth when the appropriate monitoring and controlling techniques were used.[49] Powerfull grating spectrometers with photoelectric detection and Q-switched laser systems are indispensible tools for further research near the performance limits of the method (crater diameter of 1 μm).

The other research trend favours giant single pulse lasers in the spectroscopy of highly ionized atoms, for example Ca X, Ca XVI, Fe XVII, Ni XVII and Ni XVIII, partly for classifying spectral lines which were recorded by satellites and space rockets and could not be identified.[8,30] Although these studies on plasma flares and pressure waves, which are also of great importance for nuclear fusion problems, are carried out in the vacuum ultra-violet spectrum region and with imaging optics of 1000 mm focal length, some of the results probably will in the future development of instrumentation influence the choice of technical features of the very much smaller pulsed lasers for micro-analysers.

Recent results of work on the effect of the material properties on the vaporization processes and on the plasma formation in irradiation with non-switched pulsed lasers of 10 J rating will likewise have a bearing on the further improvement of quantitative analysis.

In 1965 Felske, Hagenah and Laqua[33] reported at the International Spectroscopy Colloquium at Exeter on their efforts to obtain in the spectrochemical analysis of a somewhat larger sample with laser pulses of high repetition rate a statistical average value of high precision and accuracy. The advantages of excitation by laser radiation instead of usual spark excitation are evident: errors can be avoided, if necessary by working in the atmosphere of an inert gas[48] which are caused by undesirable diffuse forms of the discharge, without concentration in a 'focal spot' and with insufficient material removal, or by so-called mixed discharge types.

Of greatest importance are the developments in the field of sulphur, phosphorus and carbon analysis; here the requirements are very high voltages of 5 to 10 kV and external ignition of the auxiliary electrodes with a constant small electrode gap.

Even the determination of the halogens (for instance Cl III 319·14 nm[72] may become feasible.

Besides the development trends in instrumentation variations of the methods will in future probably become more prominent. Fluorescence spectroscopy with laser excitation has to be mentioned first of all. The ruby radiation of wavelength 694·3 nm is suitable for exciting fluorescence spectra of various halides, oxides and other inorganic crystals containing for instance traces of $Nd^{3+}$, $Tm^{3+}$, $Sm^{2+}$ and of other rare earths and of $U^{3+}$. In such work the monochromatic laser radiation is concentrated so that the largest possible sample volume is irradiated, and the emitted fluorescence radiation is received by a powerful photographic glass prism spectrograph at right angles to the direction of the exciting radiation. About three laser pulses are sufficient to excite fluorescence. Some cations have comparatively broad fluorescence bands ($Cr^{3+}$, $Sm^{2+}$ in corundum and fluorite), for others relatively narrow bands are characteristic ($Nd^{3+}$ in chlorides etc.). Fluorescence spectroscopy with laser excitation certainly brought an advance in certain studies in the science of mineral deposits and in regional geochemistry, especially when the analysis can be carried out at low temperatures. Finally the trace analysis of semiconductors and non-conducting materials by mass spectroscopy can be considerably improved by employing a pulsed laser (laser assisted spark source[90]). The minimum need of sample preparation and the separation of the vaporization and excitation processes are particularly stressed.

# 5

# Performance Comparison and Possible Combinations of Analytical Methods

First of all it must be made clear in which characteristic aspects laser micro-spectrochemical analysis and micro-analysis with X-ray spectrometers and electron beam excitation differ.

The phenomenon utilized in X-ray micro-emission analysis is the excitation of characteristic X-ray radiation of the elements present in a certain sample region by retardation of fast electrons in the sample material. This short-wavelength radiation has relatively few spectrum lines; it can be analysed and recorded with powerful X-ray spectrometers which are focusing instruments, not Bragg spectrometers, with curved crystals, sometimes specially cut to curvature and polished according to Johann or Johannson; the receivers are proportional flow or scintillation counters.

Castaing[22] in 1951 was the first to show how a sample area of about 1 μm diameter can be analysed by X-ray spectroscopy with an electron optic system producing narrow electron beams. His first instrument was fitted with electrostatic lenses; the next, and nowadays all commercially available efficient X-ray micro-analysers (electron probes, microprobes etc.), have an electron optic system with electromagnetic lenses (condenser and objective). Another important component of the outfit, as of the laser micro-spectrum analyser, is an optical microscope for examination in bright field illumination, sometimes with polarizing and phase contrast accessories.

The spectrometers of nearly all electron probes are reflecting instruments, semi-focusing or fully focusing, with a recording device; yet the probe designed by Borovski in the USSR has a focusing transmission spectrometer. In both types of instrument the sample can during irradiation be viewed under the microscope in reflected light.

During the primary excitation the sample as well as the microscope objective are of course in vacuum. This is a disadvantage of the X-ray microprobe compared with the laser micro-spectrum analyser in which sample and microscope are in air; apart from the need of vacuum pumps and a control system, the dimensions of the samples that can be accommodated are thereby restricted. In the *x–y* plane only samples with diameters between 0·1 cm and 2·54 cm or, in the Microscan V model of the

Cambridge Instrument Co., England, of maximum size 4·5 × 8 cm, will fit into the instrument. The maximum acceptable sample height also varies considerably from model to model, the limits being 0·5 mm in the Hitachi XMA-4 and 1·0 cm in the Microscan V.

It has to be noted that a micro-analysis of macro-scale samples is possible with a laser micro-spectrum analyser (maximum sample height for the LMA model of VEB Carl Zeiss, Jena, is 8 cm) whereas with a X-ray microprobe a micro-analysis is only possible on relatively small samples. X-ray work with fully focusing spectrometers requires considerable mechanical high-precision devices for the necessary control of movement of the analyser crystal and detector during the analysis. In laser microspectroscopy on the other hand the whole spectrograph outfit is fixed in position.

The wavelength range of laser micro-spectrochemical analysis is 200 to 1000 nm. So far the laser micro spectrum analyser has not had to be linked with a vacuum spectrograph. But a future extension of the method in this direction is to be expected to deal with special problems of the steel industry, mainly analysis of phosphorus, sulphur and carbon below 185 nm.

Elements with atomic numbers below 20 in X-ray micro-analysis emit their characteristic radiation which is already absorbed by air, hence the X-ray spectrometers of electron probes must allow evacuation. Up to 1965 even in vacuum only elements with atomic numbers between 12 (Mg) and 92 (U) could be spectroscopically determined with standard instruments because no analyser crystals with adequate reflectance for the characteristic radiation of elements with atomic numbers below 12 were available for scientific instrumentation.

To make detection of elements with lower atomic numbers by X-ray micro-analysis also possible-some workers adopted the non-dispersive method of pulse height analysis. Appropriate attachments for X-ray micro-analysers and for electron microscopes appeared (RMA unit of Siemens and Halske AG). Ranzetta and Scott of the Atomic Weapons Research Establishment, Aldermaston, England[102] developed a system for the Microscan X-ray probe of the Cambridge Instrument Co. with which they could determine carbon, nitrogen and oxygen and, by means of scan images, show the distribution of carbides and oxides in Be, Fe and U alloys.

The Siemens attachment for the Elmiskop I according to Neff allows X-ray micro-emission analysis of electron microscope samples; the facility to apply difficult techniques is a prerequisite. The system has the advantage of possible simultaneous imaging of the sample with the electron microscope by transmission, although with a reduced resolution, with a useful magnification range for 40 kV from × 70 to × 15000. On the other hand there is the disadvantage of limitation of the analytical range to elements with atomic numbers between 13 and 40; and these as a rule are only qualitatively established, normally from the K-spectra.

Fundamentally pulse height analysis is useful for elements the atomic numbers of which differ by at least three. The detector is a flow counter. As it works within the range of linear relationship, the strength of the emitted pulses is proportional to the energy of the incident X-ray quanta. The pulse height curve from which the local chemical composition within the indicated range of atomic numbers is deduced, is displayed on the oscilloscope screen of a pulse spectroscope incorporated in the radiation measuring unit.

Isolating filters have to be introduced between the exit window of the Elmiscope and the proportional counter of the RMA model in the analysis of elements that occupy neighbouring positions in the periodic table. The pulse spectrograms can be photographed with a Steinheil oscillograph camera with polaroid cassette to give a permanent record.

In 1968 the Elmisonde ('Elmiprobe') was brought out which is an Elmiskop I A attachment with two spectrometers. Non-dispersive X-ray analysis can be carried out either with the counting tube of the dispersing crystal unit in the 180° position or with two independent detectors. Back scattered and absorbed electrons and electrons emerging from the back surface of thin sample slices can also be measured.

Progress was made in the field of dispersive X-ray micro-analysis of elements with low atomic number in several instances. At the National Physical Laboratory at Teddington, England, Siegbahn gratings with 7500 lines for vacuum spectrometers were ruled allowing the resolution of the characteristic lines of C (4·4 nm), N (3·2 nm) and O (2.4 nm) and even the determination of the longer wavelength radiations of Be (11·3 nm) and Li (24 nm). Some micro-analysers were fitted with lead or barium stearate crystals for the analytical range 1·5 to 9·3 nm, such as the Philips Norelco, the Cameca, Jeol and ARL models.

The designation 'laser micro-analysis' is fully justified if one bases the comparison of the performance of important analytical methods on the sample quantity used which is ultimately decisive for the absolute limits of detection.

Wet-chemical analysis requires the digestion of a sample quantity of the order of grammes.

For a spectrochemical emission analysis milligramme quantities of material have to be prepared.

Laser micro-spectrochemical analysis without special sample preparation uses up substance quantities in the microgramme range which are characteristic of matrix and inclusion because the focused laser beam penetrates fairly deeply into solids, for instance in sulphides more than 100 μm.

The electron probe excites X-ray emission from a *surface* section in the nano-to-picogramme range which is of course not always typical of regions inside a solid.

The best limits of detection—although at the cost of resolution—are reached in non-dispersive pulse-height spectroscopy; this qualitative method is restricted to about three dozen elements and needs preparations that cannot always be undertaken. The attainable limits of detection for Cr and Fe have been reported to be about $10^{-13}$ g, corresponding to the volume of the irradiated mass of an evaporated coating of 4 nm thickness and 2 μm diameter.[120]

The relative limit of detection in X-ray micro-analysis is only in extremely favourable cases as low as 0·01 per cent (Jeol-JXA-3A model: 0·01 per cent Cu in Fe). Investigations carried out so far with laser micro-analysers have shown that in many applications a limit of 0·01 per cent is fairly easy to achieve, as for instance in determinations of heavy metals in ores, of rare earths in silicates and so on.

The advent of the solid-state laser generally improved the absolute limit of detection to values below $10^{-6}$ g; under optimum conditions the limit is as low as $10^{-11}$ g to $10^{-12}$ g (see also[103] and Chapter 4.3); the relative limit of detection is 0·01 to 0·001 per cent of the vaporized sample quantity ($10^{-6}$ g). In conventional spectrochemical analysis with spark or arc excitation the absolute limit of detection is only rarely as low as $10^{-6}$ g; the relative limit is only better by one or two orders of magnitude ($10^{-4}$ to $10^{-5}$ per cent) although a quantity of material 10000 times as big (10 mg) must be vaporized.

Scientists in the USSR[59] found that in quantitative emission spectrochemical analysis with a laser source and cross excitation (two-step method) a similar precision can be achieved as with conventional d.c. arc excitation. They also recognize the value of using the laser as an energy source, mainly because there is no need for preparation of very pure and very hard samples.

Since 1960 electron probes exist with a scanning system according to Cosslett, Duncumb and Melford; they superseded many earlier static X-ray micro-analysers.

Cosslett was the first to suggest adoption of the well-known television scanning technique for the electron probe. In this method the distribution of a certain element within a defined region of the sample surface is investigated. To do this the focused electron beam is made to sweep over the surface area in a line grid by a system of deflecting coils; the time base of an oscillograph is synchronized with this sweep. The image brightness represents the intensity of the characteristic X-ray radiation.

There are also scanning micro-analysers in which the sample in its mount moves across the fixed electron beam (Cameca probe with mechanical drive in one direction – Jeol).[57]

The electronic scanning systems of modern X-ray micro-analysers have in addition to the oscillograph displaying the characteristic X-ray image of a particular element at least one other oscillograph. It shows the relief image of the surface of the examined sample section formed by partial reflection of electrons at the sample surface. In 1964 a significant advance in producing scan images by back scattered electrons was reported from Japan; the JXA-ESA-D electronic scanning apparatus of the firm of Jeolnow contains a twin detector scanning system. It consists of two semiconductor receivers arranged symmetrically above the sample and replacing the single scintillation counter used so far. Summation of the receiver outputs gives an electron scan image the contrast of which only depends on the difference between the atomic numbers, thus giving information on the chemical composition. Output subtraction provides topographic images, likewise stereoscopic.

Not only the X-ray image of one chemical element can be presented on the JXA-3A electron probe; even the simultaneous inspection of the different X-ray images of two distinct elements is possible.

Further information on the chemical nature is provided by the scan image of absorbed electrons. For thin foils there is furthermore the electron image formed by transmitted electrons.

A few years ago two instrument types were made: purely static probes, like the first French design of Castaing, and scanning analysers, for example the Cambridge Microscan I. Now nearly all electron probes are constructed both for high sensitivity static work with high resolution and for rapid scanning of X-ray and electron images.

Laser micro-analysers and basic electron probe outfits provide information on the composition of the sample material in the micro-range; and so do scanning attachments with two to five oscillographs. Scientists working in material technology and crystal chemistry now call for instruments that in addition will also allow them to carry out X-ray crystal structure analyses in the micro-range. Between 1961 and 1963 Brümmer and others[20] drew attention to such a complex instrument constructed in Halle. With this laboratory model the following types of work can be carried out, partly simultaneously, without any changes in the set-up:

X-ray micro-analysis with primary excitation
Lattice interferences
X-ray and electron shadow microscopy
Wide angle interferences
Electron diffraction according to Lebedeff.

X-ray wide angle interferences by transmission and X-ray shadow images produced simultaneously give so called RIS (X-ray interference shadow) diagrams. Wavelength and sample thickness have to be appropriately correlated.

The optical microscope of the instrument at Halle is outside the combined electron diffraction and electron probe unit. When the sample region to be analysed has been found under the microscope in reflected or transmitted light and, if necessary, also been photographed, the test piece in its mount has to be transferred into the sample chamber of the electron optic unit. This is done by means of a motor driven carriage of a special sample manipulator fitted with a programming switch. The X-ray micro-focus procedures can be carried out either in vacuum or in air.

Nearly all the practical applications of these methods are to be found in the fields of metallography and metal physics. The reason is that recording X-ray micro-spectroscopy and its scanning methods cannot easily be adapted for work on non-conductors such as silicate minerals and glass.

Agrell and Long[2] described the most important factors that prevent the use of a scanning X-ray micro-analyser in mineralogy. It is difficult to obtain an adequate surface polish on transmitting minerals which is suitable for a large take-off angle of the X-ray radiation. Furthermore the low electrical and thermal conductivity of most minerals and rocks necessitates radiation current values two orders of magnitude below the value permissible for metals; sometimes even application of a conducting film (Al or similar material) is necessary. Scanning the surface of non-conducting materials with the focused electron beam can fairly easily produce electrostatic charges and thereby cause interference with the field of the electromagnetic lenses.

Mineralogists and geologists also want to analyse thin polished sample slices. Agrell and Long[2] found that polished disks of about 45 $\mu$m thickness can be used. Apparently induced by these somewhat successful investigations the Cambridge Instrument Company have replaced their Microscan range of instruments, well proved in metallurgical routine analysis, by the Geoscan model; this has been specially developed for extending X-ray micro-spectrochemical analysis to the fields of geology, mineralogy and petrography. The Geoscan contains two fully focusing vacuum spectrometers which allow the scanning of fairly large samples up to $4{\cdot}5 \times 8{\cdot}0 \times 1{\cdot}0$ cm. A commendable feature is the coaxial arrangement of the electron and the light optic systems; in the Microscan a rotation through 180° was necessary when the sample, aligned by light optic, had to be brought under the analysing electron beam. Similar to the Microscan is the design of the micro-analyser of Mulvey; but this has the axis of the sample drum orientated horizontally because the electron optic unit is placed under the table top of the analyser. The beams of the laser microprobes generally are coaxial. Further noteworthy new features of the Geoscan are the constant take-off angle of 75°, transistorized electronics and print-out attachment.*

In spite of appreciated innovations in the Geoscan II model of 1966 mineralogists still have to resign themselves to the impossibility of determining with an X-ray probe lithium and beryllium, two elements so very important in geochemistry.

**Translator's footnote.* The Geoscan has been replaced by the Microscan 5 X-ray micro-analyser.

TABLE 25. Important technical

| *Instrument* | *Maker* | *Probe Focal spot diameter* (μm) | *Radiation voltage* (kV) | *Scanning range* (mm) | *Sample size in x-y plane* (cm) | *Sample height* (cm) |
|---|---|---|---|---|---|---|
| SEM 2 | AEI, Manchester | 0·25–750 | 5–50 in 10 steps | 0·3 × 0·3 | 1·2 × 1·2 | 1·2 |
| Microscan V | Cambridge Scientific Instrument Co. Ltd, Cambridge | 0·3 at 25 kV | 4–50 | 1·35 × 1·35 | 4·5 × 8 | 1 |
| MS 46 DLC | Cameca, Corbevoie, Paris | 0·2–300 | 10–40 | 0·3 × 0·3 | 2·54 dia. | 1·02 |
| JXA-3A | Jeol, Tokyo | 0·3–500 | 10–50 in 10 steps | 0·33 × 0·33 | 1·2 dia. | 0·8 |
| JXA-5 | Jeol, Tokyo | 0·3–500 | 5–50 | | 26 dia. | 19 |
| XMA-4 | Hitachi Ltd, Tokyo | ≤1 | 10–50 | | 1 dia. | 0·5 |
| XMA-5 | Hitachi Ltd, Tokyo | 0·5–300 | 5–50 in 8 steps | | 25 dia. | 12 |
| 400 S | MAC, USA | ≥0·15 | 2–50 | | 3·2 and 0·4 dia. | 1·9 |
| AMR/3 | Philips, Mt. Vernon, N.Y. | 1–200 | 0–50 continuously variable | 0·053 × 0·053 to 0·32 × 0·32 | 2·5 dia. | 1·5 |

data of X-ray micro-analysers

| *Light microscope magnification* | *Atomic numbers of detectable elements* | *Spectrometer Type and numbers. Dia. of Rowland circle. Focusing. Take-off angle* | *Wavelength range* (nm) | *Special features* |
|---|---|---|---|---|
| × 100<br>× 200<br>× 400 | 5–92 | Johannson. Rowland circle diameter 50 cm | 0·05–6·7 | Provision of micro X-ray diffraction. Resolution: Electron image 0·5 μm X-ray image 1 μm |
| × 160<br>× 600 | 5–92 | Two spectrometers. Fully focusing, Take-off angle 75° | 0·1–9·1 | Horizontal column. |
| × 400 | 5–92 | Four spectrometers. Fully focusing, Take-off angle 18° | 0·07–9·3 | Receiver slit near to analyser crystal (mica and quartz). Electron optic imaging system at 50 kV. Electron probe and light microscope coaxial for simultaneous use |
| × 400 | 12–92 | Two spectrometers. Johann. $\Delta\lambda/\lambda$ $1 \times 10^{-3}$ to $5 \times 10^{-3}$ Take-off angle 20° | 0·077–1·27 | Scanning attachment with four or five oscillographs; twin detector scanning system for stereoscopic electron images |
| × 400 | 5–92 | Two (three) linear spectrometers. Fully focusing | | |
| × 400 | 12–92 | Take-off angle 15° | 0·06–1·0 | XMA-4B with soft X-ray analyser (non-dispersive for C, N, O) and 5 to 50 kV |
| × 300 | 5–92 | Three spectrometers max., linear. Fully focusing. Take-off angle 38° | | Content mapping unit |
| × 85 to × 642 | 5–92 | Three spectrometers Johanson. Fully focusing. Take-off angle 38·5° | | Development to on-line Automat |
| × 300 | < 12–92 | Johann. Take-off angle 15° | 0·14–1·4 | Range extension with Pb-stearate crystal 1·5–7 nm |

TABLE 25. (cont.)

| *Instrument* | *Maker* | *Probe Focal spot diameter* (μm) | *Radiation voltage* (kV) | *Scanning range* (mm) | *Sample size in x-y plane* (cm) | *Sample height* (cm) |
|---|---|---|---|---|---|---|
| AMX | ARL, Glendale, Calif., USA Licence: Shimadzu Seisakusho, Kyoto, Japan | 0·5–3000 | 2–30 | 0·36 × 0·36 | 15·24 dia. | 3·8 |
| EMX | ARL, Glendale, Calif., USA Licence: Shimadzu Seisakusho, Koyto, Japan | 0·3–300 | 0–50 | 0·36 × 0·36 | 15·24 dia. | 3·8 |
| TRA-25 (Tronalyzer) | Akashi Seisakusho, Tokyo | About 0·5 | 50 | 1·84 × 1·38 | 1·5 dia. | 0·8 |
| Elmisonde | Siemens & Halske, Karlsruhe | < 0·2 | 4–40 | | Up to 2·5 dia. | 0·15 |

Table 25 summarizes the important data of a number of X-ray micro-analysers. According to Mulvey[91] there are twelve manufacturers which up to September 1964 have sold about two hundred instruments throughout the world. By the end of 1966 the number had risen to six hundred.

Laser microprobes for spectrochemical analysis are so far only being offered by three instrument makers.

The given performance survey (see also)[80] establishes the fact that at present two basically different instrument types open up the field of micro-analysis: on the one hand complex outfits with electron probes, and considerable electronic equipment, evidently only practical for large research establishments; on the other hand adaptable small laser probes well suited for further build-up, which today can only be described as attachments to photographic spectrographs.

Snetsinger and Keil also arrive at the finding that 'the detection limits are such as to make the laser microprobe a useful supplement to the electron microprobe: intense continuous background (Bremsstrahlung) generated by decelerating electrons, often prevents accurate determination of elements present in concentration below about 1000 ppm with the electron probe').[121] (Cerrai and Trucco[25] express their opinion by

TABLE 25. (cont.)

| *Light microscope magnification* | *Atomic numbers of detectable elements* | *Spectrometer Type and numbers Dia. of Rowland circle. Focusing Take-of angle* | *Wavelength range* (nm) | *Special features* |
|---|---|---|---|---|
| × 100 to × 1050 | 4–92 | One to two spectrometers. Johannson. Take-off angle 52·5° | 0·1–9·3 | Spectrometer with four crystals and two detectors, Bausch & Lomb reflecting optic |
| × 100 to × 1050 | 4–92 | Three spectrometers | 0·1–9·3 | |
| | 5–92 | Take-off angle 90° | | |
| × 30 × 800 | 5–92 | Two spectrometers. Semi-focusing. Take-off angle 30° | | Attachment to Elmiskop 1A. Electron optic magnification × 70 to × 15000; one or two additional counting tubes outside the spectrometers for non-dispersive X-ray analysis |

saying that the two techniques cannot be fully equivalent substitutes for each other, since their aims may be different and their fields of application do not completely overlap.)

The following comparison, showing the difference in dimensions, may be found very instructive:

| *Instrument* | *Linear dimensions* m | *Weight* kg | *Instrument* | *Linear dimensions* m | *Weight* kg |
|---|---|---|---|---|---|
| Jeol JXA-3A | | | ZEISS LMA 1 | | |
| Electron probe | 1·3 × 1·2 × 2 | 900 | Microscope | 0·24 × 0·35 × 0·41 | 30 |
| Electronic console | 1·08 × 0·64 × 1·825 | 380 | Solid-State Laser Head | 0·14 × 0·35 × 0·15 | 5 |
| Recorder | 0·41 × 0·47 × 0·83 | 57 | Mains Supply Unit | 0·6 × 0·75 × 1·75 | 250 |
| Scanning attachment | 0·56 × 0·6 × 1·8 | 260 | Spectrograph | 1·5 × 0·4 × 0·6 | 105 |
| | | 1597 | | | 390 |

TABLE 26. Comparison of three analytical methods for examining inclusions (target) in a matrix (sample)

| | *Conventional spectrochemical analysis* | *Laser micro-spectrochemical analysis* | *Electron beam micro-analysis* |
|---|---|---|---|
| Material quantity required for analysis: | mg | μg | ng |
| Minimum size of target: | mm | 10 μm | 1 μm |
| Acceptable sample size: | | Length: little restriction<br>Width: about 100 mm<br>Height: 75 mm | Very restricted (varies) |
| Sample preparation: | Reduce size | Usually not required – Surface may be unprepared and need not be flat | Cut out required size – grind and polish surface |
| Preparation of target: | Isolate. Pulverize (risk of contamination and loss) | Not required | Not required |
| Condition after analysis: | | | |
| Sample: | Destroyed | Preserved | Partially preserved |
| Target: | Vaporized | Vaporized | Preserved |
| Examination of non-conductors: | Difficult. Hence pulverize and fill into carrier electrodes | No difficulty | Difficult: Special models of the probe, for example Geoscan. Conducting coatings required (Target then not visible). |
| Analysis: | | | |
| Qualitative | Very good | Very good | Too costly |
| Quantitative | Possible. Reference standards required | Possible. Corresponding to conventional spectroscopy with photographic recording. Homogeneous reference standard required | Possible. Very good, rapid, recording. Homogeneous reference standard required or correction by calculation |

Table 26 provides the basis for a comprehensive judgment of laser micro-spectrochemical analysis by comparing it with two others widely used analytical methods.

Different from instruments for laser spectrochemical analysis the development of commercial apparatus for micro-spark spot analysis with contact electrodes seems to be essentially completed. This is a procedure for chemical phase analysis in which, for instance, solution residues[64] and mineral grains are spectroscopically examined.

The MSL 1 model from the USSR allows the performance of local atomic spectroscopy in the spark discharge spot range from 0·08 to 0·5 mm. The required exposure times are about 10 s, that is several orders of magnitudes longer than those of laser emission spectroscopy. A special tool is supplied with the MSL 1 instrument for preparing the contact electrodes from a metal wire about 1 mm in diameter; the tips of the wire are thinned down by rolling to 0·1 mm diameter and pointed to an angle of 15–30°. A diagnostic etching with nitric acid or acqua regia follows. As the spark only penetrates into the sample material to a depth of a few micrometres this instrument from the USSR may be classified as a typical surface analyser.

Following the example of the micro-spark according to Vogel[130] a micro-spark instrument called Atomprobe was designed in the USA.[26] The improvement on the special laboratory models used before 1965[63b, 115]

TABLE 26. (cont.)

| | *Conventional spectrochemical analysis* | *Laser micro-spectrochemical analysis* | *Electron beam micro-analysis* |
|---|---|---|---|
| | A number of elements can be simultaneously analysed | | Only a few elements can be simultaneously analysed, depending on the design of probe (price!) |
| Restriction on atomic number: | No | No | Yes |
| Relative limits of detection: | 0·01–0·0001 per cent | 0·1 per cent 0·01 per cent | 1 per cent, 0·1 per cent |
| Vacuum: | For about sixty elements not required. | | Always required |
| Operation and breakdown service: | Simple | Simple | Complicated |
| Price: | Moderate | Moderate | High |

consisted in the linking of the microscope and spark source units with a powerful spectrograph *f*/6·8 for photographic or photoelectric recording of the spectra. As laser micro-spectrum analysers can also directly be fitted to grating instruments[10,49] the evident advantages of the use of the laser are fully maintained, the advantages being elimination of sample preparation, increased precision in locating the impact area (aiming accuracy) and, with the use of the masking technique, negligible oxidation.

Various combinations of operating modes are feasible in the different fields of application. Geochemists and petrographers for instance now may extend a new form of analysis for mineral determinations suggested in 1962 by H. Moenke [W. Kleber and H. Moenke, *Monatsberichte der Deutschen Akademie der Wissenschaften* (Monthly Reports of the German Academy of Science), Berlin 1964, *6*, 384]; this could be done by adopting the laser micro-method and later also by resorting to complex electron probe outfits in the following steps of operation:

Stage I: (*a*) Recording the absorption curve of the pulverized mineral in the range 4000 to 400 $cm^{-1}$

(*b*) Determination of the cation by emission spectroscopy with a solid-state laser as energy source

Stage II: Morphological crystal examination with or without polarizing microscope (checking results of analysis).

Stage III: X-ray procedures (macro- and micro-range) and Raman spectroscopy (increasing the precision of analytical findings).

The fundamental concept of method combination for mineralogists and chemists can be taken to be the sequence: infra-red spectroscopy—microscopy—emission spectrochemical analysis with laser energy source. The typical combination for metallographers on the other hand is: microscopy in reflected light with electron and laser probe.

# Bibliography


M. W. Achmanova, W. A. Kaigorodov and A. W. Karyakin: Investigation into evaporation process and plasma formation under the action of laser radiation on different substances. *J. appl. Spektr.*, 1968, **8**, 584 (in Russian).

R. V. Adams, H. Rawson, D. G. Fisher and P. Worthington: The use of the electron probe X-ray microanalyser for the indentification of inhomogeneity in glass. *Glass Technology*, 1966, **7**, 98.

AEI Elektronenstrahl-Mikrosonde SEM 2 zur Analyse von Festkörpern, Leaflet 60-04-1010 of Sunvic Regler GmbH, Solingen.

M. J. Aitken: *Physics and Archaelogy* (New York and London, 1961).

Akashi Seisakusho Ltd, Tokio: Tronscope-Tronalyzer-Prospekt.

Anon.: Development work with lasers by the Hughes Aircraft Company. *Machinery (London)*, 1962, **101**, 1310, 2612.

Anon.: Faster pulsing boosts laser applications. *Amer. Machinist Metalworking Manufact.*, 1963, **107**, 13, 60.

Anon.: Lasers boost spectrograph utility. Jarrell–Ash grafts laser energy source to standard spectrograph, comes up with microemission analysis. *Chem. Engng. News*, 1962, **40**/36, 52.

Anon.: NASA/Ames finds new extraterrestrial compounds in meteorites, *Spectrum Scanner*, 1969, **23**, 12.

Anon.: Vorschau auf die Achema-Ausstellungstagung Frankfurt/Main 1964. *Z. Instrumentenkunde*, 1964, **72**, 149.

Applied Lasers, Inc., Stoneham, Massachusetts: High repetition high energy laser system.

Applied Research Laboratories Inc., Glendale, California: The EMX-electron microprobe x-ray analyzer AMX.

E. Archbold and T. P. Hughes: Electron temperature in a laser-heated plasma. *Nature (London)*, 1964, **204**, 670.

ARL-Spectrographers Newsletter, 1960, 12/2.

ARL, Glendale: AMX-, EMX-catalogue 11/65.

Beckman & Whitley Inc., San Carlos, California: Thin film laser Q-switches for ruby lasers. *Photoinstrument Product Bulletin.*

F. Brech: Spectrochemical analysis using a laser microprobe. *Laser focus*, 1965, **1**, 5.

F. Brech: Current status and potentials of laser excited spectrochemical analysis. Publ. Jarrell-Ash Co., Waltham, USA.

F. Brech and L. Cross: Optical microemission stimulated by a ruby maser. *Appl. Spectroscopy*, 1962, **16**, 59.

J. M. Burch: Laser applications in the field of physics and optics. *Proc. Int. Symp. on Laser Physics and Applications*, 1965, **16**, 111.

Cambridge Scientific Instrument Co. Ltd., Cambridge, England: Microscan 5 X-ray micro-analysis System (catalogne).

Cameca, Courbevoie (Seine): The electronic microprobe. The new Cameca Type MS/46 electron probe microanalyser. Leaflit 1708.

Cameca, Courbevoie: Description detaillée et characteristique de la microsonde Cameca Type 46 DL. Notice Technique 1966.
R. Castaing: Mikroanalyse durch Röntgenemissions-Spektrographie. *Angew. Chem.*, 1963, **75**, 161.
R. Castaing and A. Guinier: *Proc. 1th nternat. Ionf. Con Electron Microscopy Delft* 1949. (Published by Martinus Nijhoff, Den Haag 1960), p. 60.
Compagnie Générale d'Electricité, Marcoussis: Laser catalogue.
E. S. Dayhoff: The emission mode patterns of ruby lasers. *Proc. X Coll. Spec. Inter. Washington*, 1963, p. 421.
J. Debras-Guédon: Sur l'utilisation du laser comme source d'excitation des spectres en spectrographie d'emission. *Chimie Analytique*, 1967, **49**, 419.
J. Debras-Guédon, N. Liodec and J. Vilnat: Le laser: Source d'excitation des spectres pour l'analyse qualitative et quantitative par spectrographie. *J. Mathém. Phys. Appl. (ZAMP)*, 1965, **16**, 155.
P. Dickens, P. König and P. Jaensch: Gefügeuntersuchungen von Erzen und Schlacken mit dem Mikroanalysator. *Z. analyt. Chem.*, 1965, **209**, 136.
Eastman Kodak Company, Rochester: Kodak-Laser-Glass Data Sheet.
W. Eitel: *The Physical Chemistry of the Silicates* (Chicago, Illinois, 1954).
Electro Powerpacs Inc., Cambridge, Mass.: News release 3 (1964).
R. D. Engquist: The laser matalworking tool of the future. *Machinery (New York)*, 1963, **69**, 77.
A. Engström, V. Cosslett and H. Pattee: *X-ray microscopy and X-ray microanalysis* (Proc. of the Second Intern. Symposium, Stockholm 1960) (Amsterdam, London, New York and Princeton 1960).
F. Fankhauser: Die Gefahren der Laserstrahlung. *Schweizer Archiv*, 1965, 199.
D. Feldmann: Lichttechnische Probleme mit hochenergetischen Lasern. *Lichttechnik*, 1964, **9**, 445.
A. Felske, W. D. Hagenah and K. Laqua: Laser-Anordnungen als Energiequelle zur Materialverdampfung für die Spektralanalyse. *XI Coll. Spec. Inter., Belgrade*, 1963, Abstracts (B-23).
H. I. S. Ferguson, J. E. Mentall and R. W. Nicholls: Laser excitation of powdered solids. *Nature (London)*, 1964, **204**, 1295.
D. Fishlock: The rise of the laser. *Metalworking Prod. (London)*, 1963, **107**, 74.
P. Görlich, H. Karras, G. Kötitz and R. Lehmann: Spectroscopic properties of activated laser crystals. *Physica Status solidi (Berlin)*, 1964, **5**, 437; 1964, **6**, 277.
P. Görlich and L. Moenke–Blankenburg: Laser in der Spektrometrie. *Exp. Techn. d. Physik*, 1969, XVII, 105.
H. J. Graf: Laser – Ursprung und Entwicklung, Stand der Technik, Ausblick. *Das Instr. d. Chemie*, H. 2 (1963).
W. D. Hagenah: Anwendung von Lasern bei der Materialuntersuchung. *Z. angew. Math. Physik*, 1965, **16**, 130.
W. D. Hagenah, K. Laqua and H. v. Emde: Emissionsspektralanalysen mit Materialverdampfung durch Laser-Anordnungen. *XI Coll. Spec. Inter., Belgrade*, 1963, Abstract B-24.
W. D. Hagenah, K. Laqua and H. Klocke: Materialverdampfung für die Emissionsspektralanalyse durch Laser-Anordnungen. l.c. Abstract B-25.
H. Haken: 'Laser'. *Naturwiss. Rdsch.*, 1963, **16**, 298.
T. J. Harris: High-speed photographs of laser-induced heating. *IBM-J.*, 1963, 342.
O. S. Heavens: *Optical masers* (Methuen & Co., London, 1964).

Hitachi Ltd, Tokio: Hitachi XMA-4 electron probe X-ray microanalyzer (Catalogue EX-E 038), XMA-5 (EX-E 276).
Hitachi Ltd, Tokio: Ultra soft X-ray analyzer Type YN-1. Model XMA-4B electron probe microanalyzer.
Hughes Aircraft Co., Newport Beach: Hughes ruby lasers (catalogue).
Isomet Corporation, Palisades Park, New Jersey: *Certified Threshold Lasers* (Bull. 1001).
W. F. Jankewitsch and J. W. Besrutschko: Stand for spectrochemical spot analysis. *Zavod. Lab.* 1964, **30**, 628.
Japan Electron Optics Laboratory Co. Ltd: Laser ruby rod (catalogue no. 64, 616-L 4 Tp).
Jeol, Tokio: Electron probe X-ray microanalyzer. Catalogue 6512389-E 4 Tp. JEOL Production Line Nr. 6664-Np.
A. K. Kamal: *Laser Abstracts*, Vol. 1 (Plenum Press, New York 1964) (743 abstracts of laser literature up to mid 1963).
A. V. Karyakin, M. V. Achmanova and V. A. Kaigorodov: The application of optical quantum generators in spectrochemical analysis. *Z. analyt. Chem.*, 1965, **20**, 145 (in Russian).
Y. Katsuno, H. Sunahara, K. Morita and T. Takeuchi: Laser microprobe spectrochemical analysis. *Analyt. Forschung*, 1967, **16**, 151 (in Japanese).
H. H. Klinger: *Laser – Grundlagen, Wirkungsweise und Anwendungen*, Stuttgart 1964. (In the series Radio- und Elektronik-Monographien f. d. Praktiker).
Korad Corporation, Santa Monica, California: Laser and lasers accessory equipment.
Korad Corporation, Santa Monica, California: Laser catalogues K-2Q, K-1Q, K-1.
N. W. Koroljew and G. A. Faiwilewitsch: Micro-spectrochemical analysis of nonmetallic inclusions. *Zavod. Lab*, 1964, **30**, 557.
T. Li and S. D. Sims: A calorimeter for energy measurements of optical masers. *Appl. Optics*, 1962, **1**, 325.
St. M. MacNeille: Laser beam welding – how good is it? *The Tool* and *Manufacturing Engineer*, June 1963, 59.
H. Malissa and H. H. Arlt: Elektronenstrahl-Mikroanalyse. *Radex-Rundschau*, 1964, 204.
T. H. Maiman: Optically-pumped solid-state lasers. Korad reprinted from *Solid State Design*, Nov. 1963.
R. E. Maringer, N. A. Richard and A. E. Austin: Microbeam analysis of Widmanstätten structure in meteoritic iron. *Trans. metallurg. Soc. AIME*, 1959, **215**, 56.
Maser Optics, Inc., Boston, Mass.: Laser devices, systems and accessories for 1965.
Maser Optics, Inc., Boston, Mass.: Series 600 lightweight laser system; Laser devices; High power laser family (3020-3100-3300-3750); Series MW Metal working equipment (catalogues).
J. I. Masters, P. Kafalas and E. M. E. Murray: *Bull. Amer. Phys. Soc.*, 1964, II-9, 66.
W. Meinel, R. Neubert and G. Wiederhold: Wirkungsweise der Laser und Demonstration einiger ihrer Eigenschaften. *Feingerätetechnik*, 1963, **12**, 493.
Adolf Meller Company, Rhode Island: Catalogue of optical masers.
H. Moenke: Methodical properties of laser-microspectroscopy. *Proc. Analyt. Chem. Conf., Budapest*, 1966, Vol. III, 219.
H. Moenke: Entwicklungstendenzen der Emissions-, Absorptions- und Ramananalyse mit Festkörper- und Gaslaser. *Z. Chem.*, 1966, **6**, 449.
H. Moenke: Fortschritte auf dem Gebiet der Laser-Mikrospektralanalyse. *Fortschr. Mineral.*, 1967, **44**/2, 173.

H. MOENKE: Über den Stand der Mikro-Emissions-Spektralanalyse mit Festkörperlasern. *Messtechnik*, 1968, **1**, 13.

H. MOENKE: *Mineralspektren* (Berlin 1962); *Mineralspektren*, II. Teil (Berlin 1966).

H. MOENKE and L. MOENKE–BLANKENBURG: Laser-Mikrospektralanalyse. *Jenaer Rdsch.* 1966, **11**, 166.

H. MOENKE and L. MOENKE–BLANKENBURG: Anwendungsmöglichkeiten der Laser-Mikro-Spektralanalyse in der Silikatindustrie. *Silikat Journal*, 1966, **5**, 266.

H. MOENKE and L. MOENKE-BLANKENBURG: *Optische Bestimmungsverfahren und Geräte für Mineralogen und Chemiker* (Leipzig, 1965).

L. MOENKE-BLANKENBURG: Fields of application of laser-microspectroscopy. *Proc. Analyt. Chem. Conf., Budapest*, 1966, Vol. III, 226.

L. MOENKE-BLANKENBURG: Funktion und Einsatzmöglichkeit des Laser-Mikro-Spektralanalysators LMA 1 aus JENA. *Microchim. Acta*, 1967, Suppl. II, 82.

L. MOENKE-BLANKENBURG and H. MOENKE: Der Laser-Mikrospektralanalysator – ein neues Analysengerät. *G-I-T*, 1966, **10**, 85.

L. MOENKE-BLANKENBURG and H. MOENKE: Der Laser-Mikrospektralanalysator und seine Anwendung. *Prakt. Metallographie*, 1966, **135**, 105.

L. MOENKE-BLANKENBURG and H. MOENKE: Il microanalizzatore spettrale laser e la sua applicazione. *Aluminio Nuova Metallurgia*, 1966, **35**, 441.

L. MOENKE-BLANKENBURG and H. MOENKE: Der Arbeitsbereich des Laser-Mikro-Spektralanalysators LMA 1. *Bergakademie*, 1966, **18**, 697.

H. NEFF: Über die Röntgen-Emissionsanalyse von elektronenmikroskopischen Präparaten. *Z. Instrumentenkunde*, 1964, **72**, 125.

N. NIKOLOV, A. P. PETKOV and G. DIMITROV: Laser microspectral investigations of meteorites. *XV Coll. Spec. Int., Madrid*, 1969, Abstract 407.

I. NODDACK and W. NODDACK: Herkunftsuntersuchungen. *Angew. Chemie*, 1934, **47**, 637.

J. NORRIS: Some applications of a laser microprobe for the generation of emission spectra in organic materials. Paper read at Spring Technical Conference on the Applications of Lasers and Masers to Science, Medicine and Industry JEEE, Cincinnati, Ohio, 7 April 1964.

R. D. OLT: Synthetic Maser Ruby. *Appl. Optics*, 1962, **1**, 25.

A. E. PALADINO: Preparation of crystal and glass laser material. Part I. *Semiconductor Products*, 1964, **7**, 31.

A. PETRAKIEV and G. DIMITROV: Laser spectral analysis with utilization of time resolved spectra. *XV Coll. Spec. Int., Madrid*, 1969, Abstract 409.

Philips Electronic Instruments, Mount Vernon, N.Y.: Norelco electron probe microanalyzer (Leaflet RC 399-Rev. 363-2 MF).

N. B. PIPER: Aufbauvorschriften für optische Laser-Rubinstäbe. *Microtechnic (Lausanne)*, 1963, **17**, 74.

S. P. S. PORTO: The ruby optical maser as a Raman source. *Proc. X Coll. Spec. Inter., Washington*, 1963, p. 417.

Rank Precision Industries Ltd (Analytical Division, Hilger & Watts): Photographic spectrographs (catalogue CH 403/8, 1970).

Raytheon, Waltham, Massachusetts: Lasers and accessories (catalogue 7–64). Electronic process 7/4 (1963).

J. F. READY: Development of plume material vaporized by giant-pulse-laser. *Appl. Phys Letters*, 1963, **3**, 11.

S. L. RIDGWAY, G. L. CLARK and C. M. YORK: Time-resolved spectroscopy of ruby laser emissions. *J. O. S. A.*, 1963, **53**, 700.

H. Riesenberg: Ein neues Mikroskop-Spiegelobjektiv mit großem Arbeitsabstand. *Jenaer Rdsch.*, Messesonderheft, 1964, 40.
E. F. Runge, F. R. Bryan and R. W. Minck: Laser-excitation of atomic spectra. *Canad. Spectroscopy*, 1964, **9**/5.
N. M. Saks and C. A. Roth: Ruby laser as a microsurgical instrument. *Science (Washington)*, 1963, **141**.
K. Shimoda: Threshold condition of masers for higher frequencies. *Appl. Optics*, 1962, **1**, 303.
Siemens & Halske A.G., Munich: Solid state lasers.
G. F. Smith: Applications of lasers to instrumentation. *ISA Transactions*, 1964, **3**, 353.
H. Statz: Optical masers. *Semiconductor Products*, 1962, p. 17.
S. Sugano: Spectroscopy for solid state optical masers. *Appl. Optics*, 1962, **1**, 295.
P. A. Thiessen, K. Meyer and W. Meier: Strukturveränderungen in Antimonkristallen durch punktförmige mechanische Bearbeitung und Laserbestrahlung. *Monatsber. DAdW*, 1965, **7**, 391.
V. T. Tomberg: Non-thermal biological effects of laser beams. *Nature (London)*, 1964, **204**, 868.
C. H. Townes: Optical masers and their possible applications to biology. *Biophys. J.*, 1962, **2**, 325.
G. Troup: *Masers and Lasers* (Methuen & Co., London and John Wiley & Sons, New York, 1963).
Union Carbide, Linde Division, New York: 2122 Linde Laser Welder (catalogue F-2122 4-64 8M-0366).
Valpey Crystal Corporation, Holliston, Mass.: Laser-Optics (catalogue).
H. J. F. van Draat: Microanalyse met lasers. *Chemie u. Technik Revue*, 1967, **22**, 133.
VEB Carl Zeiss, Jena: Laser-Mikrospektralanalysator LMA 1. Leaflet W 32-372-1. Instruction manual no. 32-G 372a-1.
K. Vogel und P. Backlund: Application of electron and optical microscopy in studying laser-irradiated metal surfaces. *J. Applied Phys.*, 1965, **36**, 3697.
H. Westmeyer: Zur Optik für Laserstrahlbearbeitungsgeräte. *Feingerätetechn.*, 1965, **14**, 529.
J. P. Wilson: Fortschritte auf dem Laser-Gebiet in England. *Int. Elektron. Rdsch.*, 1964, 621.

## REFERENCES CITED

(The bold numbers in the square brackets refer to the pages where the references are cited.)

1. M. W. Achmanova, W. A. Kaigorodov and A. W. Karyakin: Examination of laser-produced condensed plasmas by infra-red spectroscopy. *J. appl. Spektr.*, 1968, **8**, 858 (in Russian). **[221]**
2. S. O. Agrell and J. V. P. Long: The application of the scanning X-ray microanalyser to mineralogy. In A. Engström, V. Cosslett and H. Pattee: *X-ray microscopy and X-ray microanalysis* (Proc. of the Second Internat. Symposium, Stockholm 1960), Amsterdam, London, New York and Princeton, 1960, p. 391. **[229]**
3. Ch. Allemand: Application du laser à la spectroscopie d'emission, *Metallurgia (ital.)*, 1964, **56**, 411. **[26]**
4. American Optical Company, Southbridge: AOLux neodymium-doped glass laser rods. Catalogue 13-5180 A. **[4]**

5. ANON.: Bestrahlung von Zellen mit Hilfe eines Rubin-Lasers. *Atompraxis*, 1963, **9**, 158. [**12, 32, 101, 102**]
6. ANON.: New application of laser microprobe, *JEOL News*, 1967, **5**, No. 9, p. 8. [**28**]
7. E. ARCHBOLD, D. W. HARPER and T. P. HUGHES: Time resolved spectroscopy of laser-generated microplasmas. *Brit. J. appl. Phys.*, 1964, **15**, 1321. [**29, 55, 75, 92, 201**]
8. N. G. BASOW, V. A. BOJKO, JU. P. BOJNOW, Z. R. KONONOW, O. N. KROCHUN, S. P. MANDELSTAM and G. V. ZKLUEKOW: Laser flash for exciting spectra of highley ionizee atoms. *Rev. Roum. Phys.*, 1968, **13**, 97. [**222**]
9. G. BAUDIN and C. BLAIN: Application du Laser a l'analyse locale. *XV Coll. Spec. Int., Madrid*, 1969, Abstract 302. [**28, 181, 221**]
10. E. S. BEATRICE, I. HARDING-BARLOW and D. GLICK: Laser-microprobe emission spectroscopic analysis of elements in single cells. *Lowry-Cocroft Laser Abstract*, 1968, No. 404. [**236**]
11. E. S. BEATRICE, I. HARDING-BARLOW and D. GLICK: Electric spark cross-excitation in laser microprobe-emission spectroscopy for samples of 10–25μ diameter. *Appl. Spectr.*, 1969, **23**, 257. [**117**]
12. M. BERNDT, W. GRASSME, E. O. KOCH and W. MEINEL: Laser-Geräte und Laser-Resonatoren aus Jena. *Jenaer Rdsch.*, Sonderheft, 1964. 3. [**18**]
13. M. BERNDT, H. KRAUSE, L. MOENKE-BLANKENBURG and H. MOENKE: Mikro-Emissionsspektralanalyse mit Zeiss-Festkörperlasern. *Jenaer Jb.*, 1965, **I**, 45. [**19, 58, 65, 75, 103, 133**]
14. M. BESSIS, F. GIRES, G. MAYER and G. NOMARSKI: Irradiation des organites cellulaires à l'aide d'un Laser à rubis. *C. R. Séances Acad. Sci. Paris*, 1962, **255**, 1010. [**31**]
15. W. H. BLACKBURN, Y. J. A. PELLETIER and W. H. DENNEN: Spectrochemical determinations in garnets using a laser microprobe. *Appl. Spectr.*, 1968, **22**, 278. [**148, 219**]
16. H. J. BLANKENBURG, H. MOENKE, L. MOENKE-BLANKENBURG and K. WEHRBERGER: Einsatz von Laser-Mikrospektralanalyse und UR 10-Mikroeinrichtung bei der Bestimmung kristalliner Einschlüsse in Bergkristallen. *Kristall u. Technik*, 1966, **1**, 351. [**144, 167**]
17. F. BRECH and J. B. MARLING: Laser excited spectrochemical analysis. *XI Coll. Spec. Int.*, Belgrade, 1963, Abstract P 4 (Unija Hemijskih Društava SFR Yugoslavia). [**132**]
18. M. BRUMA and M. VELGHE: Adaption d'un dispositif laser sur un microscope. Photo-63269 of C. N. R. S. Paris. [**45**]
19. M. BRUMA and M. VELGHE: Experience montrant la materialisation du pointage optique par fasceau laser. Micropointure laser. Photo-63275 of C. N. R. S. Paris. [**45**]
20. O. BRÜMMER, K. H. BRAUER and G. SUWALKI: Über ein Gerät zur wahlweisen Verwendung als Röntgenmikroanalysator, Röntgenschattenmikroskop und Elektronenbeugungsanlage und über experimentelle Untersuchungen an Al–Cu und an Ge. *Z. angew. Physik*, 1963, **16**, 27. [**228**]
21. J. M. BURCH: Optical research and the solid state laser. *J. sci. Instruments*, 1963, **40**, 147. [**99**]
22. R. CASTAING: Dissertation Univ. de Paris 1951; Publ. O. N. E. R. A. Nr. 55. [**212, 224**]
23. E. CERRAI and R. TRUCCO: Alcune esperienze sull'applicazione della Campionatura mediante laser alla spettroscopia di emissione; determinazione di elementi, effetti della struttura dei materiali. *Metallurgia (ital.)*, 1967, **59**, 615. [**195**]

24. E. Cerrai and R. Trucco: Laser microprobe sampling for qualitative analysis. *Energia nucleare*, 1968, **15**, 342. [**159, 163, 192, 197**]
25. E. Cerrai and R. Trucco: On the 'matrix effect' in laser sampled spectrochemical analysis. *Energia nucleare*, 1968, **15**, 581. [**23, 151, 195, 204, 212, 232**]
26. G. Chaplenko, D. O. Landon and A. J. Mitteldorf: The atomprobe – a progress report. *Proc. XIII Coll. Spec. Int., Debrecen*, 1967. [**235**]
27. J. Debras-Guédon and N. Liodec: De l'utilisation du fasceau issu d'un amplificateur à ondes lumineuses par emission induite de rayonnement (Laser à rubis), comme source ènergétique pour l'excitation des spectres d'émission des elements. *C. R. Acad. Sci. (Paris)*, 1964, **257**, 3336. [**71, 72, 90, 187, 192, 198, 200**]
28. J. Debras-Guédon and N. Liodec: Sur une extension des possibilités de l'analyse spectrale dans le domaine de l'analyse ponctuelle par utilisation d'un laser. *Bull. Soc. franç. céram.*, 1963, **61**, 61. [**71, 90, 198**]
29. F. Dörr: Zur Anwendung der Elektronenmikroanalysatoren in den Glaslaboratorien. *Glastechn. Ber.*, 1966, **39**, 141. [**166**]
30. B. C. Fawcett, A. H. Gabriel, F. E. Irons, N. J. Peacock and P. A. H. Saunders: Extreme ultra-violet spectra from laser-produced plasmas. *Proc. Phys. Soc.*, 1966, **88**, 1051. [**222**]
31. A. Felske, W. D. Hagenah, H. Klocke and K. Laqua: Fortschritte in der spektrochemischen Lokalanalyse mit Lasern. Paper read at the conference on Optics and Spectroscopy of all Wavelength Regions, Jena 1965. [**95**]
32. A. Felske, W. D. Hagenah and K. Laqua: Spektrochemische Analyse mit Laser-Impulsen höherer Folgefrequenz. *Z. analyt. Chem.*, 1966, **216**, 50. [**94, 205, 207, 208, 209, 210**]
33. A. Felske, W. D. Hagenah and K. Laqua: Spektrochemische Analyse mit Laser-Impulsen hoher Folgefrequenz. *Proc. XII Coll. Spec. Int. Exeter*, 1965, p. 340 (Adam Hilger Ltd, London). [**223**]
34. A. G. Fox and Li Tingye: *Bell System techn. J.*, 1961, **40**, 453. [**5**]
35. H. Freund (Hrsg.): *Handbuch der Mikroskopie in der Technik*, Bd. IV, Teil 4: Mikroskopische Forschung in der Glas- und Emailtechnik, Frankfurt 1964. [**170**]
36. F. Früngel and H. G. Patzke: Laser für die Werkstoffbearbeitung. *Z. Ver. dtsch. Ing.*, 1964, **106**, 787. [**9**]
37. N. M. Galdina, J. S. Janowskij, N. G. Kuznecova and M. A. Babdshanjan: Der The new electricallx fused high-resistance stone Bbkor for glass furnaces. *Steklo i. Keram.*, 1962, **19**, 15. [**171**]
38. E. Gegus: Eine spektrochemische Lösungsmethode zur Analyse von Einschlüssen in Stählen, in R. Ritschl and G. Holdt: *Emissionsspektroskopie* (Berlin 1964) p. 306. [**154**]
39. A. D. Genkin and N. W. Koroljew: Method of determining small mineral quantites in ores. Геология Рудных месторождений *(Geology of Ore Deposits)*, 1961, No. 5, 64. [**129**]
40. L. Georgieva, A. Petrakiev and G. Dimitrov: Laser microspectral analysis of galenite and pyrite in a magnetic field. *Scient. Progr. 3rd Bulg. Conf. on Spectr. Varna* (1968), 20. [**137**]
41. Wa. Gerlach and We. Gerlach: Über die Anwendung der emissionsspektralanalytischen Methode bei Aufgaben der sozialen Medizin. *Angew. Chemie*, 1934, **47**, 825. [**179**]
42. W. Gerlach: Methoden und Anwendungen der Emissions-Spektralanalyse in der medizinischen Praxis und der medizinisch-biologischen Forschung. *Zeiss-Nachr.*, 1939, **3**, 2. [**179**]

43. R. Geyer, K. Doerffel and H. Kirst: Ein besonders einfaches Verfahren der spektrochemischen Lokalanalyse. *Z. anal. Chem.*, 1960, **172**, 326. **[173]**
44. D. Glick: The laser microprobe–its use for elemental analysis in histochemistry. *J. Histochem. a. Cytochem.*, 1966, **14**, 862. **[179]**
45. D. Glick and R. C. Rosan: Laser microprobe for elemental microanalysis–application in histochemistry. *Microchem. J. New York*, 1966, **10**, 393. **[179]**
46. H. M. Goldman, M. P. Ruben and D. Sherman: The application of laser spectroscopy for the qualitative and quantitative analysis of inorganic components of calcified tissues. *Oral surgery*, 1964, **17**, 102. **[179]**
47. K. Gürs and H. Westermeier: Ein leistungsfähiger Rubinlaser von besonders kleinen Abmessungen. *Frequenz* 1964 **18**/7. **[220]**
48. W. D. Hagenah, K. Hirokawa and K. Laqua: Über die Eigenschaften von verdampfenden und nicht verdampfenden Entladungsformen bei Funkenanregung in Edelgasen. *Proc. XII Coll. Spec. Int., Exeter*, 1965, p. 334 (Adam Hilger Ltd). **[223]**
49. I. Harding-Barlow, E. S. Beatrice and D. Glick: Laser microprobe emission spectroscopy of picogram quantities of cations in tissues. *Lowry–Cocroft Laser Abstract*, 1967, No. 1230. **[222, 227, 236]**
50. P. Hein: Ein Mikroobjektiv mit großem Objektabstand und hoher Apertur. *Feingerätetechnik*, 1965, **14**, 133. **[51]**
51. Jarrell-Ash Co., Waltham, USA: *Laser microprobe mark II, Model 45-604.* Technical Bulletin SP 1 (1966). **[46, 70, 150, 179, 221]**
52. Jarrell-Ash Co., Waltham, USA: *Spectrum Scanner*, 1966, **21**, No. 3 and 4. [173, 176, 179]
53. Jarrell-Ash Co., Waltham, USA: *Laser microprobe mark II model 45-604* (Instruction manual). **[46, 70]**
54. Jarrell-Ash Newsletter No. 14 (1963) 6. **[25]**
55. Jarrell-Ash Co., Newtonville, USA: Bulletin 45-600; Bulletin 45-601, Cat.-No. 3.3. **[25, 33, 136]**
56. Jarrell-Ash Co., Waltham, USA; Jarrell-Ash S. A. Cat. Nr. 1.2 von W. Zeh, Duisburg: 1,5 m stigmatischer Spektrograph (20. 3. 1964). **[25]**
57. Jeol, Tokio: *Electron probe X-ray microanalyzer JXA-3A* (Catalogue no. 649355-E 5Tp). **[227]**
58. A. V. Karyakin, M. V. Achmanova and V. A. Kaigorodov: Two-step method of spectrum excitation. *J. angew. Spektroskopie*, 1965, **2**, 364 (in Russian). **[118, 212]**
59. A. V. Karyakin, M. V. Achmanova and V. A. Kaigorodov: Possibilities for the use of a laser in the spectrochemical analysis of pure substances. *Proc. XII Coll. Spec. Int., Exeter*, 1965, p. 353 (Adam Hilger Ltd). **[227]**
60. Y. Katsuno, Morinaka and H. Sunahara: Emission spectrochemical analysis with laser radiation. *Japan. Analyst*, 1968, **17**, 376 (in Japanese). **[28, 156, 179, 205, 206, 207, 215]**
61. P. H. Kim and S. Namba: Beam angle and output energy of ruby laser in external mirror alignment. *Japan. J. appl. Phys.*, 1964, **3**, 305. **[14]**
62. P. Klinger and W. Koch: *Beiträge zur metallkundlichen Analyse*, Teile 1 and 3, Düsseldorf 1949. **[154, 170]**
63a. F. A. Koroljew: Generators and amplifiers of coherent light and prospects of their application in spectrum analysis. *Proc. 16th Spectroscopy Conference of the U.S.S.R., Moscow*, 1965. **[84, 92, 94, 139, 157]**
Further contributions from:
N. S. Sswentizki; N.K. Ossipov, K. I. Taganov and L. M. Feinberg; N. K. Ossipov, N. S. Sswentizki and S. I. Schlepkova; L. N. Kaporski, N. K. Ossipov and

N. S. SSWENTIZKI; L. N. KAPORSKI and S. I. SCHLEPKOVA; G. I. BAKANOWITSCH, L. I. GRETSCHICHIN and L. J. MINKO; A. V. KARJAKIN, M. V. ACHMANOVA and V. A. KAIGORODOV; W. W. PANTELEJEV and A. A. JANKOWSKI.

63b. N. W. KOROLJEW: Outfit for emission micro-spectrochemical analysis. Заводская Лаборатория *(Betriebslaboratorium)*, 1959, **25**, 1014 (in Russian). **[130,235]**

64. N. V. KOROLEV, V. V. RJUCHIN and M. J. KOLPAKOV: A Russian instrument for optical emission micro-spectroscopy – an appendix to the Russian edition of this book (published by MIR, Moscow 1968). **[118, 235]**

65. J. KROSS: Monochromat 1:1,5 für eine Sonderaufgabe. *Mitt. Ber. Opt. Inst. TU Berlin*, 1966/10 1966, 10 4. **[53]**

66a. M. LANDWEHR: Die wissenschaftlichen Arbeiten zur Bekämpfung der Staublungenerkrankung (Silikose) der Bergleute und Forschungen über die Entstehung der Silikose im Bergbau, in H. FREUND: *Handbuch d. Mikroskopie in der Technik.* Bd. II, Teil 2, Frankfurt a. M. 1964, p. 597. **[180]**

66b. K. LENZ: Erfahrungen mit Rubinlasern. *Mber. dtsch. Akad. Wiss. Berlin*, 1963, **5**, 259. **[95]**

67. CH. LÖFFLER and J. LÖFFLER: Über die Bestimmung der Tonerde in Schlieren mit Hilfe der Elektronenstrahlmikrosonde. *Glastechn. Ber.*, 1966, **39**, 333. **[166]**

68. J. R. LOOFBOUROW: Microspectroscopy. *J. opt. Soc. Amer.*, 1950, **40**, 317. **[129]**

69a. M. LOQUIN: Lasers et microscopes. *Z. wiss. Mikroskopie*, 1964, **60**, 68. **[43]**

69b. K. F. LÜDEMANN, W. SCHIRMER and R. EBERT: Ergebnisse der Untersuchung einiger Schlacken- und Eisenfunde von Berlin-Köpenick. *Schr. Sekt. Vor- u. Frühgesch. dtsch. Akad. Wiss. Berlin*, 1962, **12**, 107. **[174]**

70. T. H. MAIMAN: Stimulated optical radiation in ruby. *Nature (London)*, 1960, **187**, 493. **[1, 3]**

71. H. MALISSA: *Einführung in die Elektronenstrahl-Mikroanalyse* (Wien 1966). **[205]**

72. Publications of the Russian scientific-Technologieal Conference on the Appliea-tion of Spectrum Analysis in Industry and Scientific Research (Tin Russian) Contributions of N. W. KOROLJEW, W. W. RJUCHIN, L. P. RATTIAN 27–28, 55–59; Minsk 1967. **[26, 42, 129, 223]**

73. J. MATSUSHITA, Y. TUNEKO and S. SUZUKI: An experiment in laser microspectrochemical analysis. Abstracts of a meeting on Analytical Measuring Methods in Short-time Spectroscopy, Tokyo 1969. Abs. 20. **[187, 198, 199, 215]**

74. J. A. MAXWELL: The Laser – as a source in emission spectroscopy. *Chem. in Canada*, 1963, **15**, 10. **[78, 99, 117, 132]**

75. J. A. MAXWELL: The laser as a tool in mineral identification. *Canad. Mineralogist*, 1963, **7**, 727. **[78, 117, 133, 197, 200, 202]**

76. F. J. MCCLUNG and R. W. HELLWARTH: Giant optical pulsation from ruby. *J. Appl. Phys.*, 1962, **33**, 828. **[15]**

77. R. C. MELLORS and K. G. CARROLL: A new method for local chemical analysis of human tissue. *Nature (London)*, 1961, **192**, 1090. **[181]**

78. H. U. MÖDE, W. D. HAGENAH and K. LAQUA: Verlängerung und Verstärkung der spektralen Emission von Analysenproben, die durch Laser-Licht verdampft worden sind. *XV Coll. Spec. Int., Madrid*, 1969, Abstract 288. **[222]**

79. H. MOENKE: *Spektralanalyse von Mineralien und Gesteinen* (Leipzig 1962). **[128, 135, 267]**

80. H. MOENKE: Vergleichende Betrachtung über Elektronenstrahl-Mikroanalyse und Laser-Mikrospektralanalyse. *Microchim. Acta*, 1967, Suppl. II, 93 and 105. **[232]**

81. H. MOENKE: Moderne mineralogisch-geochemische Untersuchungsmethodik mit

neuartigen Analysengeräten. *Ber. dt. ges. Geol. Wiss. B. Miner. Lagerstättenf.*, 1968, **13**, 323. [**145**]

82. H. Moenke and L. Moenke-Blankenburg: Laser-Mikrospektralanalyse. *Prakt. Chem. (Wien)*, 1965, **16**, 225. [**19**]
83. H. Moenke and L. Moenke-Blankenburg: Der Laser-Mikro-Spektralanalysator und sein Einsatz in der Glas-, Keramik- und Zementindustrie. *SPRECHSAAL für Keramik, Glas, Email, Silikate*, 1967, **100**, 112. [**164, 172**]
84. H. Moenke and L. Moenke-Blankenburg: *Introduction to laser micro-emission spectroscopy* (Published by MIR, Moscow, 1968) (in Russian). [**vii**]
85. H. Moenke, L. Moenke-Blankenburg and W. Quillfeldt: Laser-Mikro-Emissionsspektralanalyse in verschiedenen Wellenlängenbereichen. *Mikrochimica Acta*, 1968, Suppl. III, 221. [**56, 61**]
86. L. Moenke-Blankenburg: Geochemische Spurenanalyse mit Gitterspektrographen, in R. Ritschl and G. Holdt: *Emissionsspektroskopie* (Akademie-Verlag, Berlin 1964) p. 351. [**81, 140**]
87. L. Moenke-Blankenburg: Laser-Mikro-Spektralanalysator LMA 1 aus Jena. *Feingerätetechnik*, 1968, **17**, 193. [**56**]
88. L. Moenke-Blankenburg and J. Mohr: Laser-Mikrospektralanalyse mit gütegesteuertem Laser. *Jenaer Jahrbuch* 1970, 195. [**14, 125, 138, 155, 163**]
89. P. L. Morris, D. F. Askey and W. M. Pardue: Laser microprobe as an analytical tool in plutonium research. *Lowry-Cocroft Laser Abstract*, 1968, No. 1063. [**222**]
90. V. G. Mossotti, D. M. Golightly and W. C. Phillips: Laser-assisted spark source mass spectrographic trace analysis of semiconducting and nonconducting materials. *XV Coll. Spec. Int., Madrid*, 1969, Abstract 310. [**223**]
91. T. Mulvey: First Colloquium on electron probe X-ray microanalysis, Delft, September 1964 and *J. Sci. Instruments*, 1965, **42**, 57. Further more: Programme of First European Colloquium on Electron Probe X-ray Microanalysis. [**232**]
92. S. Namba and P. H. Kim: Electron and laser beam procressing. Japan. *J. appl. Phys.* 1964, **3**, 536. [**14**]
93. H. Neuninger: Personal communication 1969. [**183, 184, 215**]
94. Optics Technology Inc., Belmont, Cal.: Model 190 Microscope Laser (Catalogue 21364) and Laboratory Laser Model 120 (Catalogue 52962). [**29, 103**]
95. H. Otto: Die Anwendung der Spektralanalyse auf kulturhistorische Fragen. *Spektrochim. Acta (Berlin)*, 1940, **1**, 381. [**175**]
96. W. W. Panteleev and A. A. Jankovskij: Applications in spectroscopy of optical quantum generators working in single pulse mode. *J. appl. Spektr.*, 1968, **8**, 905 (in Russian). [**56, 91**]
97. N. A. Peppers: A laser microscope. *Appl. Optics*, 1965, **4**, 555. [**29, 30**]
98. A. Petrakiev, G. Dimitrov, S. Beltschew and N. Nikolow: Mikrospektraluntersuchungen des Meteoriten 'Pawel'. *Jena Review*, 1970, H. 1, 21. [**134**]
99. A. Petrakiev, A. Samow and G. Dimitrov: Die Anwendung des Laser-Mikrospektralanalysators LMA-1 des VEB Carl Zeiss, Jena, für die Identifizierung von Pigmenten in mehrschichtigen Gemälden alter Wandmalereien. *Jena Review*, 1971, p. 250. [**176, 177, 178**]
100. R. Pittioni: Zweck und Ziel spektralanalytischer Untersuchungen für die Urgeschichte des Kupferschieferbergwesens. *Archaeologia Austriaca*, 1959, **26**, 67. [**173**]
101. W. Quillfeldt: Über optimale Anregungsbedingungen bei der Lasermikrospektralanalyse mit Fremdzündung der Hilfsfunkenanregung. *Exper. Techn. d. Phys.*, 1969, **XVII**, H. 5, 415. [**64, 65, 66, 67, 68, 69**]

102. G.V.T.Ranzetta and V.D.Scott: Electron-probe microanalysis of low atomic number elements. *Brit. J. appl. Phys.*, 1964, **15**, 263. [**225**]
103. S.D.Rasberry, B.F.Scribner and M.Margoshes: Characteristics of the laser probe for spectrochemical analysis. *Proc. XII Coll. Spec. Int., Exeter*, 1965, p.336 (Adam Hilger Ltd, London). [**26, 50, 95, 109, 116, 179, 206**]
104. S.D.Rasberry, B.F.Scribner and M.Margoshes: Laser Probe Excitation in Spectrochemical Analysis. I: Characteristics of the Source. *Applied Optics*, 1967, **6**, 81–86. [**25, 51, 70, 74, 104, 215**]
105. S.D.Rasberry, B.F.Scribner and M.Margoshes: Laser Probe Excitation in Spectrochemical Analysis. II: Investigation of Quantitative Aspects. *Applied Optics*, 1967, **6**, 87–93. [**25, 51, 70, 156, 207**]
106. R.C.Rosan: On the preparation of samples for laser microprobe analysis. *Appl. Spectroscopy*, 1965, **19**, 97. [**117, 178, 205**]
107. R.C.Rosan, F.Brech and D.Glick: Current problems in laser microprobe analysis. *Federation Proc.*, 1965, **24**, Suppl. 14, 126. [**179**]
108. R.C.Rosan, M.K.Healy and W.F.McNary: Spectroscopic ultramicroanalysis with a laser. *Science (Washington)*, 1963, **42**, 236. [**178**]
109. D.Röss: *Laseroszillatoren* (Akad. Verlagsgesellschaft Frankfurt/Main, 1966). [**1, 16**]
110. J.Rubeská and A.Spackova: Spektrographische Lokalmikroanalyse von mineralogischen Schliffen. *Časopsis Mineralog. Geolog.*, 1962, **7**, 158 (zech.). [**128, 129**]
111. E.Rudolph: Nachweis vom Fremdmetallabrieb auf Stahlwerkzeugen mit Hilfe der Laser-Mikrospektralanalyse. Thesis (not published), Köthen, 1969. [**184**]
112. E.F.Runge, S.Bonfiglio and F.R.Bryan: Spectrochemical analysis of molten metal using a pulsed laser source. *Spectrochim. Acta*, 1966, **22**, 1678. [**90**]
113. E.F.Runge, R.W.Minck and F.R.Bryan: Spectrochemical analysis using a pulsed laser source. *Spectrochim. Acta (London)*, 1964, 733. [**29, 73, 78, 90, 195**]
114. J.R.Ryan, E.Ruh and C.B.Clark: Laser microprobe analysis of glass defects. *Amer. Ceram. Soc. Bull.* 1966, **45**, 260. [**166, 211**]
115. T.N.Schadlun: Über kombinierte röntgenometrische und mikrospektralanalytische Bestimmung kleinster Körner von Erzmineralien. *N. Jb. Miner. Abh.*, 1965, **102**, 115. [**235**]
116. A.L.Schawlow and C.H.Townes: Infrared and optical masers. *Phys. Rev.*, 1958, **112**, 1940. [**1**]
117. H.Schumann: *Metallographie*, 3.Aufl., VEB Deutscher Verlag f. Grundstoffindustrie, Leipzig 1960. [**148, 152, 153**]
118. R.H.Scott and A.Strasheim: Laser induced plasmas for analytical spectroscopy. *XV Coll. Spec. Int., Madrid*, 1969, Abstract 318. [**56, 91**]
119. D.B.Sherman, M.P.Ruben, H.M.Goldman and F.Brech: The application of laser for the spectrochemical analysis of calcified tissues. *Ann. New York Acad. Sci.*, 1965, **122**, 767. [**179**]
120. Siemens & Halske AG, Karlsruhe: Einrichtung für Röntgenmikroanalyse (RMA) zum Elmiskop IA and Elmiskop I (Catalogue Eg 1/5, July 1964). [**227**]
121. K.G.Snetsinger and K.Keil: Microspectrochemical analysis of minerals with the laser microprobe. *American Mineralogist*, 1967, **52**, 1842. [**207, 215, 232**]
122. D.Socher: Laser. *Wissensch. Fortschr.*, 1964, **14**, 342. [**6**]
123. B.H.Soffer and R.H.Hoskins: Generation of giant pulses from a neodymium laser by a reversibly bleachable absorber. *Nature (London)*, 1964, **204**, 276. [**16**]

124. B. P. Stoicheff: Spectroscopic studies with the ruby optical maser. *Proc. X Coll. Spectr. Internat.*, 1962, p. 399. (Washington Spartan books 1963). **[94]**
125. D. J. Swaine: The laser as a source of excitation in emission spectroscopy. *Hilger Journal*, 1968, **11**, 69. **[117]**
126. R. Tlalka: Mikroanalyse der Kernmateriale mit der Laser-Mikrosonde. *XV Coll. Spec. Int., Madrid*, 1969, Abstract 356. **[222]**
127. TRG Inc., Syosset, N.Y.: '103-Mini-Laser; 104 Laser-System' (Catalogue). **[29]**
128. R. Trucco: A laser microprobe for spectrographic analysis. *Energia nucleare* 1968, **15**, 277. **[19, 25, 103]**
129a. VEB Carl Zeiss, Jena: Catalognes: 'Mikroskope Ng und Nf' 30-036b-1; 'UV-Spectrograph Q 24' 32-G 360c-1; Atlas 'Q 24-Eisenspektrum' from 4555 Å to 2227 Å with indicatee analytical lines of 62 elements, 2. Aufl., Leipzig 1959; 'L-Generatoren, LASER' 40-633-1 and 40-640-1; 'Festkörperlaser ZFL 750-T5' 40-G633-1. **[18, 101, 124, 135]**
129b. J. Vilnat, N. Liodec and J. Debras-Guedon: Le laser, source d'excitation des spectres: applications aux dosages de chrome et de nickel dans des aciers et à l'analyse qualitative de matières diverses. (Applications of a laser to the determination of chromium and nickel in steels and to the qualitative analysis of various materials). *Proc. XII Coll. Spec. Int., Exeter*, 1965, p. 343 (Adam Hilger Ltd, London). **[195, 212]**
130. R. C. Vogel: *Process Dev. Quarterly Report* MCW-1479, USAEC, Washington D.C. Nov. 1, 1962. l.c.: G. H. Morrison: *Trace analysis*. (New York 1965) 207. **[235]**
131. P. Volkmann: Spektralanalytische Untersuchungen an Stählen mit dem Laser-Mikrospektralanalysator. (Deutsche Akademie d. Wiss. zu Berlin 1969), Thesis (manuscript). **[104, 106, 150, 156]**
132. A. B. Whitehead and H. H. Heady: Laser-spark excitation of homogeneous powdered materials. *Appl. Spectr.* 1968, **22**, 7. **[173, 215]**
133. R. Wilson, L. Goldman and F. Brech: Calcinoiss cutis with laser microprobe analysis. *Arch. Derm.*, 1967, **95**, 491. **[179]**
134. K. Wohlleben, H. Woelk and K. Konopicky: Untersuchung kugeliger Einschlüsse im Flachglas mit Hilfe der Elektronenstrahlmikrosonde. *Glastechn. Ber.*, 1966, **39**, 329. **[166]**
135. L. Moenke-Blankenburg, H. Moenke, K. Wiegand, W. Quillfeldt, J. Mohr, W. Grassme and W. Schrön: Neue Aspekte der Gerätetechnik und der Methodik zur Laser-Mikro-Emissionsanalyse. LMA 10, ein neuer Laser-Mikro-Analysator. *XVII Coll. Spec. Int., Firenze*, 1973, Acta, Vol. II, p. 485. **[221]**

# Index